D0053253

CLOUDMONEY

ALSO BY BRETT SCOTT:

The Heretic's Guide to Global Finance

CLOUDMONEY

———

CASH, CARDS, CRYPTO,
AND THE WAR FOR OUR WALLETS

———

BRETT SCOTT

HARPER
BUSINESS
An Imprint of HarperCollins*Publishers*

CLOUDMONEY. Copyright © 2022 by Brett Scott. All rights reserved. Printed in Canada. No part of this book may be used or reproduced in any manner whatsoever without written permission except in the case of brief quotations embodied in critical articles and reviews. For information, address HarperCollins Publishers, 195 Broadway, New York, NY 10007.

HarperCollins books may be purchased for educational, business, or sales promotional use. For information, please email the Special Markets Department at SPsales@harpercollins.com.

Originally published in the United Kingdom in 2022 by The Bodley Head, an imprint of Vintage.

FIRST U.S. EDITION

Illustration credits: p.19: picture alliance / dpa | Boris Roessler; p.137: Barclays Group Archives; p.217: Ian Lambot; p.228: Pixabay.

Library of Congress Cataloging-in-Publication Data has been applied for.

ISBN 978-0-06-293631-8

22 23 24 25 26   FB   10 9 8 7 6 5 4 3 2 1

For Mum, Dad, Craig and Ant,
with much love

Contents

Introduction

This book is about a merger and an acquisition. The merger is between the forces of Big Finance and those of Big Tech. The acquisition is of power: once the merger is complete, Big Finance and Big Tech will have power over us on a scale never before seen in human history.

The argument I will make is a contrarian one. Every day the media buzzes with excitable stories about how this or that start-up company offers convenient benefits to people through this or that exciting new fintech app. When, for example, Amazon announces a new partnership with a payments platform, or Citigroup announces a collaboration with Google Pay, it's presented – and reported on – as a welcome, cutting-edge innovation. Futurists clamour to make their voices heard, weighing in on the latest buzzwords in digital finance like bards competing to sing the praises of the king.

I want to show you why you should distrust narratives about the supposedly inevitable progress of digital money and finance. This will involve tuning out of the day-to-day chatter of the finance and tech industry, and ignoring the tales told by company CEOs and their acolytes. Entrepreneurs, like surfers, enjoy telling exciting stories about the waves they ride (while giving tips on how to stay upright), but are less interested in meditating on the confluence of hidden forces, like winds out at sea and coral reefs, that produce big swells. The swells might be the result of a distant earthquake, which is itself the result of unseen plate tectonics. I'd like us to bypass the

1

wave-riding stories and go straight to uncovering the plate tectonics of the global economy.

We are witnessing the automation of global finance, a process that first requires the physical money in our wallets to be replaced with digital money controlled by the banking sector – a scenario euphemistically called 'the cashless society'. The financial industry – and some governments – have been making a concerted effort to demonise physical money for at least the last two decades. The Covid-19 pandemic has supercharged this rhetoric, and finance and technology corporations have taken the opportunity to accelerate their war on cash, using concerns about hygiene to push this agenda even further. Cash protects privacy, and it is resilient in the face of both natural disasters and banking failures, but it is increasingly being presented as an outdated barrier to progress, and one which must inevitably give way to a new world of digital money, or what I call 'cloudmoney'.

The digitisation of payments enables the digitisation of finance more broadly – a task currently outsourced to the financial technology, or 'fintech', industry – which in turn is enabling the full automation of corporate capitalism. We can already see this at work in the operations of corporations like Amazon, Uber and Google (or, in China, Tencent and Alibaba). Almost every big tech firm is entering into partnerships with financial institutions. Such companies cannot have global digital empires without fusing with global digital payment systems.

Major oligopolies (conglomerations of mega-firms) are forming out of this process, but they are concealed behind a proliferation of apps that give the superficial appearance of diversity. Behind our smartphone screens is growing an infrastructure of automated financial control. Billions of people are being locked into interconnected systems that enable hitherto unimaginable levels of surveillance and data extraction, and bring with them grave new potential for exclusion, manipulation and delusion. The fight to lock people into dependence on these systems is becoming a geopolitical struggle between major powers, supported by their corporate allies.

On first look, individual corporates and governments would appear to be competing for dominance, but closer consideration reveals that they are elbowing each other for position within a growing planetary-scale super-system. It's difficult to glimpse this super-system in its entirety, partly because it is too large to see. But our routine interactions with a constellation of phones, computers and sensors (all of which route information to faraway datacentres) leave an imprint on us, translating into an uncomfortable sense of living in a world destined to move towards ever greater levels of concentrated interconnection.

Some, myself included, feel claustrophobia when we notice this tightening web. I shudder at adverts that showcase the convenience of products that will later attempt to study and steer my behaviour. I look at my phone and wonder if, rather than a helpful companion, it is an agent for a sinister force, monitoring aspects of my life previously resistant to formal control.

I am not here, however, to argue that the digital world is bad, or that it should be contrasted with some good non-digital world. Public debates are framed as battles between one thing and another, but I see the world more as contradictions. I recognise that we are all caught in complex webs – economic, cultural and political – that can simultaneously liberate and imprison us. This book is designed to *rebalance* the skewed digital finance story, which speaks only of liberation. Think of it as a darker yin to contrast with a brighter yang.

The contradictions of money and tech

When my brother and I were young, my dad taught us to read topographic maps and sent us off to navigate South Africa's Drakensberg

Mountains with only a compass. We thought this made us real tough guys, but in these same mountains, 500 years earlier, the indigenous Sān people had been doing just the same without any technology at all – navigating solely from experience, the stars and their intuition.

Herein lies a contradiction. A tool is – at face value – an instrument we use to impose our will upon the world, like the precious compass my brother and I held tightly as we strode forth. And sure enough, we arrived at the mountain cave before nightfall, proud of our achievement. What we struggle to see, though, is that the tool only functions at the expense of us becoming dependent on it. In using it we outsource, forget about and potentially lose part of our inner compass, or never let that inner compass develop at all.

Technology is double-sided. We experience it as empowering, and yet it increases our dependency. The external gadgets that support us come to mould our actions and thoughts, as innovations that are initially received as new options but go on to become mandatory necessities. If you live in a major city, you might be able to choose your brand of smartphone, but you cannot really 'choose' whether to use a phone. You had better choose one, or else face exclusion by the socio-economic network that surrounds you, and upon which you depend.

And this contradiction intensifies when we do not even *hold* the powerful technologies on which we are dependent. Google Maps, for example, does not exist on my phone. It resides in a distant digital datacentre – part of what is colloquially nicknamed 'the cloud' – which I access via my smartphone. It is an outsourced sense of direction, entrusted to a very large entity far away.

Our dependency on Google Maps is recent – a matter of decades. But today a Londoner's heart is set racing when their phone battery drops to 1 per cent and the prospect of losing access to that distantly controlled digital oracle looms. We build our lives around these technologies in such a way that they seem to fuse into us. Losing access to my phone for a day makes me feel like a chain-smoker on a long-distance flight, obsessing about the moment when I can disembark, rush outside, light up and feel whole again.

This pattern of contradiction exists with *money* too, but at an even deeper level. Nowadays we experience money as vital and – if you are at the right end of the income ladder – empowering, and we have long forgotten what the world looked like prior to monetary exchange, to the point where we cannot even really imagine it. Five thousand years ago, monetary systems were small and isolated, but have come to engulf our civilisation. Almost every object around us – from our computers to our shoes, and from our imported pasta to this book – is accessed via a global monetary exchange system. Our relationship with money goes even deeper than our attachment to technology, with panic setting in when our balance approaches zero and we face the prospect of losing market access. Losing that access, even worse than being a chain smoker on a flight, makes me feel like a fish slowly suffocating on dry land, heaving myself towards water. This is because money facilitates access to all the other things we are dependent upon, making it an ultimate object of dependence.

This takes on a whole new dimension, however, when we lose the ability to directly hold our money. The digital money in our bank accounts resides in distant datacentres controlled by the banking sector, which we communicate with via our phones, computers or payment cards. The 'cashless society' is a world in which our ability to transact is outsourced to these financial entities, currently entering into synergistic deals with corporations like Google, who host our ability to navigate. It is these synergies that are presented as being fantastically convenient, but this convenience comes paired with an extremely high level of dependence on concentrated corporate power. This is a core contradiction of our times, and one that I will guide you through in this book.

My journey

For the last fourteen years I have been on the front lines of the global financial sector. It began when I joined a rogue financial start-up in London, where I tried to broker obscure bets known as 'exotic derivatives' in the midst of the 2008 financial crisis. For two years, my days were spent phoning up the finance directors of corporations, the managers of mega-sized funds and traders at investment banks. The company eventually failed, buckling under the strain of the convulsing global markets, but survived long enough to give me an introduction to the dark arts of high finance.

The financial sector is an ancient power that pre-dates the Internet by at least a thousand years. It presides over the global monetary system, upon which hundreds of millions of economic interactions and financial contracts rest daily. It is the world of central banks, commercial banks, Wall Street, the City of London and the global network of onshore and offshore financial centres. The packed pubs of London's financial districts are frequented by rowdy suited employees of the sector: the loud-mouthed trader, the slick investment banker, the refined wealth adviser, the gruff hedge fund manager. Behind discreet doors in upmarket Mayfair, Russian oligarchs raise money for mining ventures while Middle Eastern oil sheikhs get investment advice for their sovereign wealth funds.

In 2013 I published a book about that world called *The Heretic's Guide to Global Finance: Hacking the Future of Money*. I have a background in anthropology, and I used approaches from that discipline to delve into Big Finance. The book also drew upon hacker philosophy – which explores how to infiltrate complex systems – as an approach to challenging the power of the financial sector. Its publication paved the way for me to criss-cross the world, collaborating with many different communities who claim to have uncovered the keys to financial revolution or reform.

These communities, from far-left anarchists and ecological activists to New Age spiritualists, market libertarians, hawkish conservatives and government technocrats, have many contrasting viewpoints. I have designed local currencies with hippies, helped climate campaigners lobby pension funds, assisted accountants in re-imagining the future of auditing, and challenged monetary policymakers. I've been a financial artist-in-residence at a Vienna art gallery and a collaborator with the MIT Media Lab. The world I inhabit includes Malaysian central bankers and American IMF officials, German anti-surveillance activists and Serbian political dissidents. I've even had occasion to sit at the same tables as far-right activists, some of whom flirt with fascism. I am fortunate in having been exposed to contrasting perspectives on the problems of our economic system, different approaches on how to change it and different end objectives in the people calling for those changes.

By 2015, I had begun to shift my focus to the lords of digital automation – the technology companies of Silicon Valley. Unlike the regimented offices of corporate finance, this is a world of open-plan workspaces with beanbag chairs, Post-It notes on whiteboards, and colourful computer code on black screens. It is the world of start-up pitches jam-packed with the language of innovation, in which CEOs walk on stage wearing a headset microphone to introduce their latest app to evangelical applause. The biggest companies – Google (Alphabet), Facebook, Apple, Amazon and Microsoft (and in China, Alibaba, Tencent and Baidu) – are embedding themselves as the connective digital tissue through which we all engage with the market. The position enables them to harvest enormous amounts of data, which they then use to train their artificial intelligences – AIs.

The upper echelons of both the financial and digital technology sectors are staffed by people who believe that their actions lead the world, but who come from contrasting cultures. While the image of the financial world is one of ruthless self-interest, as exemplified by the character of the corporate raider Gordon Gekko in *Wall Street* (1987), the image of tech has long been one of idealistic and geeky coders. Apple's iconic 1984 Superbowl advert represented this spirit – featuring a colourfully dressed athlete smashing a grey and oppressive status quo with a sledgehammer, it promised freedom from traditional power structures.

That was the 1980s, though. Now these cultures are merging. One illustration of this is the migration of staff between Big Finance and Big Tech – for instance, a friend of mine used to work at J. P. Morgan as a quantitative analyst, where he calculated the prices of financial contracts. Now he works at DeepMind, Google's AI research unit, investigating how to create AI that can be applied to any situation.

This blending of finance and tech is also visible in the hybridisation of the two industries in the realm of fintech. It is an industry that exemplifies the ambiguous yet close relationship between the two worlds. Banks had a toxic reputation in the aftermath of the 2008 financial crisis, and a techno-utopian idea emerged that digital start-ups would disrupt finance and produce financial democratisation. Digital tech was presented as a white knight that would kick bad old finance into shape. 'Fintech' became a buzzword, attracting both workers from traditional banks who had ideas about how to digitise services, and entrepreneurial technologists who wanted to take on finance as outsiders.

From the outset, fintech felt drabber than the rest of the technology sector, given that it was attached to the old power of Big Finance, but more colourful than traditional finance, given its association with the hype of Silicon Valley. To this day, it trades on the idea that technology is revamping finance, and that banks are being dragged kicking and screaming into the new digital world. In the language of tech, the future must be brought forward into the present, and everything old must disappear into the past. The old financial system must be brought up to date and the old ways – including bank branches, physical cash and non-digital processes – must die. These ideas are presented as a fundamental disruption of finance, but when I take a step back and consider the fintech industry, I see not an attempt to redesign Big Finance, but rather an attempt to *automate* it. This distinction is seldom made, though. Why is that?

The 'inevitable progress' of automation

People intuitively make predictions about how the future *could* be, but if we feel uncertain about what we see in that future, we can turn to arguments for how it *should* be. This is the field of politics, where we make impassioned pleas for the future we want, rather than settling on a future we deem likely. But while local political demands get made for, say, schools to be funded or local green infrastructure to be built, transnational demands are far harder. When it comes to the broad trajectory of digitisation and automation in the global economy, people are strangely mute. There is a sense that they will happen regardless of whether you want it.

Why? Well, our transnational economic system dwarfs each one of us, and most individuals experience it as something they learn to respond to rather than actively shape. No person feels themselves 'driving' the global economy, but they do experience it moving, like a vast procession on autopilot. It's par for the course, along with corporations getting bigger, weapons becoming more powerful, resources depleting, and digital connectivity getting ever denser.

9

This seems eerily on course with scenarios imagined by cyberpunk science fiction writers from the 1970s. Their characters live in high-tech worlds where forests have been obliterated by sprawling megalopolises, and where governments have fused with huge corporations. The latter offer dazed humans the chance to plug into virtual-reality dreamscapes to escape the treadmill of their lives, while small bands of rebels attempt to resist.

If it sometimes feels as though dystopian science fiction has inspired technology companies, it is because we already see its plot lines manifesting in real-world innovations brought to us by Big Tech: from *Minority Report*'s ubiquitous facial-recognition technology and *Blade Runner*'s biotechnology through to *Snow Crash*'s 'Gargoyles' – individuals rigged up with devices that feed audio-visual data into a virtual reality version of the Internet called the 'Metaverse'. But nobody needs to be 'inspired' by science fiction for its plot lines to play out: cyberpunk was simply extrapolating from trends that were already inherent within large-scale capitalist systems, which is why the results continue to turn up in our present, as if governed by inertia.

Because it temporarily destabilised that feeling of inertia, the Covid-19 pandemic was a profound mental jolt for many of us. Our systems appeared, briefly, to *pause* – creating anxiety in some and euphoria in others – before cranking back into the same old patterns, like a treadmill restarting (and doing so at a faster speed). Techno-optimists work hard to put a positive spin on this feeling of inertia. They argue that the ever-increasing scale and speed of economic processes is 'progress' driven by all of us, animated by the creative human spirit.

These stories pervade digital finance: for example, pundits claim that a cashless society is inevitable, because 'we' – the members of the public – see the value in ever-increasing speed, automation, connectivity and convenience, and want ever more digital finance. Because 'we' all want this, no individual dissenter can stand against it, and if they try, they will be *left behind*. This messaging is reinforced by an entire marketing industry that specialises in telling us to *get ready* for the change we are apparently driving, lest we are bypassed by a 'rapidly changing world'. This messaging accompanies almost all products pushed out by finance and tech companies, presenting commercial interests as natural forces, unstoppable and benevolent for all.

This is what I see on the platform in the London Underground, in the form of an advert for digital payment that proclaims that 'The Future is Here'. I also see it on the side of a Singapore skyscraper, as a billboard for Samsung smartphones announces that 'Next is Now'. I see it as I watch an entrepreneur on a conference stage in Nur-Sultan amen, Kazakhstan, prophesying the coming digital transformation of – everything. The same message comes out of the mouth of a local politician on TV in South Africa – my home country – as he tells us to prepare for the 'Fourth Industrial Revolution'. My dad is a former soldier from rural Zimbabwe, and uses a twelve-year-old computer, but this background noise coming out of the TV tells him to prepare for a vast complex of drones, robotics, smart cities, biotechnology and AI: all things he has never asked for.

But from where does my local politician get his message? The official story propagates from high-tech centres in powerful regions where huge amounts of profit are at stake. One such centre is 16,000 kilometres away from South Africa, in Silicon Valley, where people are raising money from investors and planning marketing campaigns to get us hooked on their platforms. Their whispers percolate from the boardrooms and bars of the Bay Area to innovation journalists, who influence panel organisers at the World Economic Forum in Davos, which gets reported on by a BBC broadcast watched by a local trendsetter in Johannesburg, who is tasked with keeping my politician up to date on international trends. This, alongside a thousand other channels, is how the technological

11

mantras of our time spread into my dad's lounge. Having been told to prepare, he will, like most people, simply experience the technologies spreading through his peer networks, after which he will have little choice but to join.

For many people there is little feeling of either collective or individual agency to choose how this unfolds. Some, however, find it psychologically easier to become a cheerleader for the coming wonders of technological progress, and to glaze over with fatalistic indifference at any mention of dystopian potentials. It helps if you get paid to do this, and many mainstream futurists get paid a *lot* of money to style themselves as prophets of inevitabilities. For example, in 2016 Kevin Kelly, the founding editor of *Wired*, published *The Inevitable: Understanding the 12 Technological Forces That Will Shape Our Future*. The title presents the future like the weather – as something that will simply happen to you. The twelfth prediction in his 'weather forecast' is that we will be absorbed into a 'planetary system connecting all humans and machines into a global matrix'.

Let me offer a suggestion for how to create this global matrix. Take an oligopolistic sector of tech giants, whose platforms are fused into the life of billions, and glue them via fintech infrastructure to an oligopolistic sector of financial giants, whose digital money is fused into the life of billions. Then glue both to everything else (cities, machines, our bodies), and present this situation – in which our entire environment is possessed by the profit motives of distant oligopolies – as an inevitable and welcome revolution driven by us all. Finally, cast anyone who rebels as an irrelevant and out-of-touch Luddite stuck in the past, who need to be cajoled along, or rescued.

The crypto wildcard

Perhaps, however, there are other ways to create a global matrix. One such proposal arrived in 2008, in the form of an obscure nine-page PDF document posted onto an Internet forum. The document was titled 'Bitcoin: A Peer-to-Peer Electronic Cash System', and it was authored by an unknown person going under the pseudonym of Satoshi Nakamoto. The paper described how a network of people could issue digital tokens and move them between each other without the involvement of banks, who preside over the normal digital money system that we use when tapping our contactless payment cards. Nakamoto and various collaborators set about building the proposed system, and by 2009 had released the first version of an open-source protocol that – when used by people – gave rise to Bitcoin, the world's first 'cryptocurrency'.

I began to experiment with Bitcoin in 2011, and wrote two initial blog posts about it then, one of which soon appeared on the first page of Google's search results for the phenomenon. When producers at the BBC and other media outlets began frantically searching for information about Bitcoin around 2013, I started receiving emails asking me to talk about it on TV and radio. I had also begun to earn Bitcoin tokens – mainly by exchanging copies of my first book for them – and had used them to get things like pizza from a pub in London, mint tea in Bulgaria, and even goods from an adult site called Crypto Sex Toys. I convinced my housemate to accept my rent in them when I ran out of normal money, and I paid helpers with it. A crypto scene began to develop off the back of Bitcoin, with new cryptocurrencies emerging. It was fun and experimental, an ethos that was exemplified by the arrival of the playful Dogecoin in 2013, a cryptocurrency based on a Shiba Inu dog meme.

The atmosphere soon changed. Speculators, intrigued by the

technological novelty of these crypto-tokens, began to pile in and trade them. At the same time, the underlying blockchain technology that underpinned these tokens was foregrounded, and by 2015 blockchain became a buzzword in its own right, excitedly touted by innovation pundits. Blockchain technology is used to create digital systems that can co-ordinate action between people who do not know each other, without the need for an intermediary. This could include moving tokens around (which is what the Bitcoin system facilitates), but could also extend beyond that. The wide range of unexplored possibilities made it a powerful catalyst for new technological visions, all based on the concept of 'decentralisation': any existing system that was 'centralised' – which means a system with small numbers of large players at its core – was seen to be under threat of disruption. That could include the financial system, but also the legal, copyright or global trade system.

While this was exciting, the vagueness of the proposed solutions, compounded with poor understanding of our existing systems, produced some outrageous assertions about how blockchain would revolutionise money, finance and economies. It was promoted by everyone from intellectual property lawyers to anarcho-capitalist libertarians, and from neo-fascists to New Age yogis, who saw in it an organic vision for global harmony.

The sheer weight of the hype brought it to the attention of mainstream institutions, a development which appeared in my inbox as emails with requests for help, media appearances and speaking invitations. I wrote one of the first United Nations reports on cryptocurrency, and later presented on it at the European Union Commission and Parliament, while IMF officials emailed me to ask whether it could be used to solve problems in the international payments system. The blockchain wave took me all over the world, from Amsterdam to San Francisco, and from Nairobi to Tokyo.

The strange truth was that I knew little about blockchain

technology, but neither did anyone else. The scene was crammed with chancers reciting catchy soundbites in the studios of Bloomberg and CNBC, or on the stages of global conferences. I have watched entrepreneurs with no experience of the complex history of colonialism argue that blockchain will 'end African poverty', and I have seen countless cryptocurrency gurus predict the demise of the banking sector without understanding how banks operate. I've also met senior bankers who take them seriously because they do not have the skills to assess the claims made by technologists.

Blockchain technology originally promised to provide a decentralised alternative to the growing finance and tech oligopolies that I alluded to at the beginning of this introduction. Its early development was directly inspired by concerns about the surveillance implications of a cashless society, and by the potential for the massive centralisation of state and corporate power in the digital age. However, blockchain technology possesses deeply ambiguous contradictions of its own. One of these is that, far from being repelled by it, financial institutions and mega-corporations seem increasingly eager to incorporate it into their operations. The same technology that can co-ordinate networks of ordinary people can be repurposed to coordinate oligopolies.

By 2021, blockchain hype had hit a new fever pitch, as the global capitalist system began to swallow whole sections of it. Tech titans like Elon Musk began promoting crypto-tokens, venture capitalists set up funds to invest in crypto start-ups, and massive global payments companies like Visa started offering new business lines to integrate crypto into normal payments systems. It might have begun as an imagined antithesis to Big Finance and Big Tech, but in reality a synthesis is emerging, and one that is just as likely to further dystopian trends as it is to combat them.

Where are we going?

Surely there must be a silver lining to the direction we're being steered in? Well, there might be, but before I can get to that, we need to take a tour of our monetary system, to understand better how it is changing and to describe the erosion of the cash system. I will then delve into the dynamics of fintech, how it tries to 're-skin' the existing financial system, and how that intersects with Silicon Valley. Next I will lead you through the often perplexing world of cryptocurrency and blockchain technology that pitches itself as an alternative. I will show the zones of hybridisation that are occurring as banks raid the crypto world, and vice versa. The narrative will lead us to the present day, where these forces now stand poised to enclose us, unless we find the strength to push them in a new direction.

In the course of this journey I will be criticising many institutions, from states and corporations, through to start-ups and even ideological communities. I wish to make clear that this is not intended as a critique of *people* within these systems. All of us have to survive in this world, and for the majority that means having to work within its existing structures. I often see those structures as having a logic that transcends the individual good intentions of those who find themselves employed by them, or even of those running them. Before we can hope to creatively reimagine our systems, we must critically introspect on them, and now is the best time to be doing this. The pandemic has consolidated our dependence upon transnational digital infrastructures, and many of us, stuck at home behind screens, have sensed not only the emptiness that lies within that enclosure, but also the hidden power that thrives there.

1

The Nervous System

I find myself looking out from the thirty-ninth-floor window of the second highest skyscraper in the UK. They call this place Level 39. It is a hub for financial technology start-ups in Canary Wharf, the London district that hosts one of the world's greatest concentrations of financial mega-corporations. Level 39 was created by the Canary Wharf Group, which owns the entire district, to grow these fintechs in a Petri dish. There are over a hundred of them in here, most working on some aspect of financial automation, from payments apps and insurance bots to AI credit-scoring and 'robo-advisers'.

These spaces full of young companies are called 'incubators' or 'accelerators', but a more accurate image might be a luxury fitness centre, where a company goes for an intensive workout session, gets pumped with steroids (venture capital support), and finishes off with an hour on a sunbed for that healthy glow. I often get invited into this tech start-up world. I am in Level 39 taking part in a workshop on 'the future of money'.

But this is not the first time I have looked out of skyscraper windows in Canary Wharf. The first time was more than a decade ago, in July 2008, when I found myself at a job interview on the thirty-fifth floor of the offices of an investment bank called Lehman

Brothers. I made it to a second interview, but before I could get to a third the mega-bank collapsed into bankruptcy and set off a global financial crisis.

As that crisis escalated, I was employed as a derivatives broker, a position that saw me visit many of these skyscraper offices. During that time I learned that the taller your building is, the less down-to-earth a mindset you must have. Nobody uses a thirty-fifth floor office to hand-make bread from stone-ground flour, for example, but they will use it to write mega-bets on the global price of wheat, to be used for transnational wheat risk-management, or for speculation.

London is not the only place where these skyscrapers rise. They rise anywhere the lords of finance gather, whether it be Singapore, New York, Shanghai, Tokyo or Frankfurt. One of the most iconic skyscrapers in Frankfurt is the Commerzbank Tower, and I remember snapping a shot of it late one night as a security guard watched

me from inside. The immense tower brought to mind the fortress of the sorcerer Saruman from *The Lord of the Rings*, its sheer wall ascending to a rooftop citadel shrouded in ghostly yellow from powerful spotlights. Every element of these buildings – from the security gates to the one-way glass panes that glint in the sun – is designed to exude a sense of impenetrable power. The architecture echoes our relationship with high finance; most people stand beneath these monoliths, on the outside looking up.

But, inside, the Commerzbank Tower has a secret: a men's toilet in which a line of ceramic urinals are positioned to offer a panoramic view over the city, so that those inside can gaze down on the people going about their lives as they take a piss.

A fitting image of condescending bankers looking down on the world while metaphorically urinating on it? Having worked in high finance, I think the picture is more complex. Beneath their bravado, bankers are seldom in control of their own institutions, and are often channelling a logic that transcends them. There is something inhuman about a corporate skyscraper. The suits bankers wear are like protective uniforms, and the toilet is in fact the only

place in a skyscraper where they might reveal a chink in that armour, baring their bum in the cubicle and revealing a warm body.

All of us, in the end, are local and communal creatures, and even the highest-flying bankers would lose the will to enter these cold towers if they had no friends, family, pets or community to return to each evening. Nobody wants to cuddle up in bed in the Commerz-bank Tower, and you cannot smell or hear the distant activities you see from its fiftieth floor. Skyscrapers are not a natural habitat for warm-blooded humans. They are, however, a natural habitat for *corporations*, if we were to conceive of those as self-contained living entities. Corporations feel very much at home in steel towers, perceiving the people seen from their fiftieth floor as mere data points to process through spreadsheets.

Viewed collectively, the global finance corporation community is like a dense nerve centre for a multi-layered empire of money – and promises for money – communicated via fibre optic cables under seabeds and routed via offshore centres to other faraway clusters of corporations. Level 39 is hosted in the top reaches of one of these towers and, while they may not know it, the fintech-industry employees within have been hired to automate this nerve centre.

Money as a nervous system

I use the term 'nerve centre' deliberately. It is common for economists to use blood metaphors when speaking about money, characterising it as a substance of value that 'flows' around the economy. Financiers love this metaphor because it presents their sector as the 'beating heart' of the global economy. But to use this circulatory system metaphor prevents us from seeing the true nature of finance.

In a human body, the nervous system is a network of neurons embedded in all tissues and muscles, via which impulses are transmitted to activate those muscles. It concentrates in certain places like our spinal cord and brain (which has the highest level of neural density). Similarly, our global monetary systems are interconnected – albeit largely invisible – networks that spread to the furthest reaches of the planet and, just as neurons are embedded in tissues, so they are embedded in *us*. But, while it reaches the dustiest small town, this system of money concentrates in the world of high finance, which itself concentrates in these high towers.

It is not surprising that many people get flustered when they hear media pundits talk about the activities that take place here. They hear statements like 'Trillions fly through the foreign exchange market daily,' or, 'The value of the global derivatives market is ten times the size of world GDP,' and so on. These descriptions evoke images of an alien world of massive-scale numbers. High finance feels ever present and yet ever divorced from our lives, but the complex networks of finance can always – in the end – be traced back to our bodies, and to the earth.

Indeed, everything in the final analysis is derived from our ecological systems, without which we perish. If I walk to a park, lie on the grass and look down, I will see mites crawling through soil particles, and if I could sink deeper I would see microbes, fungi and molecules of water. This is the substratum that gives birth to the plants, which give life to countless creatures, all arranged in ecosystems that have allowed us to survive for over two hundred thousand years, raising children, forming communities and producing goods and services, or *value*.

For a large part of those hundreds of thousands of years we were able to survive without money. What we nowadays refer to as 'the economy', however, is an interdependent network of people and groups of people who produce goods and services for each other using this ecological base, but who co-ordinate the transfer of their labour (and products of their labour) via a monetary system, which holds them together in a dense constellation. That's why from the

21

outset it is useful to see a unit of money as something that can *activate* someone into producing value – like an impulse flying from one part of an interconnected nervous system to another, resulting in movement in the body.

In subsequent chapters we'll delve deeper into the monetary system and consider how it bifurcates into separate parts, one centred on central banks and another on commercial banks (key protagonists in the battle between cash and digital money). We will be learning that cash is a more fundamental form of money, and that the digital money you use when tapping your payments card derives its power in part by tethering itself to an association with cash.

For the rest of this chapter, however, let's take for granted that the monetary web is embedded in our lives, so that we can explore how this allows the flashier players in finance (those investment banks and hedge funds) to design and trade contracts that steer, scale up and increase the complexity of monetary movements, via corporate structures. This will make clear how the high-level structure of the economy relates to the seemingly low-level question of what form of money we carry in our day-to-day lives.

Charging up corporate capitalism

Titans of industry have no interest in doing hard labour. If a team of five oil barons wishes to develop a new offshore drilling operation, the oil barons are not going to build it themselves by welding the rig together out on the open ocean. Rather, they will get *other people* to build it, mobilising an army of suppliers to provide them with materials and enlisting the services of engineers and many other contractors and labourers. The primary method of setting this process in motion is to establish a legal entity – a company or corporation – and give it a name (e.g. DeepFuel Inc.) and a bank account, so that it can do deals with these contractors and labourers. That bank account, however, needs to be 'charged up' with money (or 'capitalised') before the company can act in the world.

The financial sector is the place where that 'charging up' gets arranged. Our oil barons might transfer some of their own money to the account of DeepFuel, but the task of fully charging up the new entity will be outsourced to investment bankers. The latter will draw up an investment prospectus for DeepFuel, aiming to showcase how the managers of the operation will, subject to certain risks, be able to source the inputs they need (workers, materials, technology, etc.) for less money than the outputs (in this case, oil) can be sold for. This DeepFuel prospectus, with its tantalising promise of future profit, can then be dangled in front of the manager of a large pension fund looking for investment opportunities.

A pension fund is an institution that has rallied together money from thousands of individuals into a huge 'battery' waiting to charge up corporations seeking financing. Our investment bankers tap into this, inviting the fund managers to give DeepFuel the money it needs now in exchange for a cut of its future profit. This promise for a claim on future money is encoded in a legal contract called a share, but the company can also be charged up by promising other investors (called creditors) fixed cuts of future money, in exchange for their present money.

The true lifeblood of an economy is not money but people carrying out labour. But a heart can be made to beat through an electrical shock. Once a corporation is charged up through capitalisation, it can blast that charge out through the monetary nervous system like a defibrillator kickstarting thousands of human bodies into large-scale action. This is how ten thousand labourers can be mobilised to make and assemble the components of an oil rig, and then operate it to extract the oil which their bosses can sell to customers. Those customers are a source of uncertainty because they could be poached by a rival corporation (and so managers seek to suppress costs, for example by replacing human labour with machine labour), but as the product is sold, it sends a flash of money back up the circuit. Some of that money exits in the form of bonuses to management and tax to government, while the rest gets sent to recharge the batteries by giving investors the future money promised in their financial contracts. In this way interest payments accrue to creditors, while dividend payments accrue to shareholders.

This is how the thoughts of five oil barons get translated into action via money pushed into labour and technology markets via financial markets, eventually manifesting as products in commodity markets. If they establish their DeepFuel venture as a promising operation they might then sell it to the oil giant BP, allowing it to be subsumed as a *subsidiary*. In 2014 I worked with the Berlin-based open data company OpenOil to map the byzantine corporate structure of BP: money from big investors enters the BP mothership (via the London Stock Exchange), which in turn capitalises thirty-five sub-companies, which in turn capitalise hundreds of sub-sub-companies. The overall structure is twelve tiers deep, and what we call 'BP' is actually a federation of over 1,100 subsidiaries connected by a complex web of financial circuitry spread across the globe. This is how a mega-corporation comes to be created.

These mega-corporations control their subsidiaries from headquarters (in a skyscraper somewhere), and make them 'do business with each other', which is why up to half of global trade actually takes place *within* corporations, rather than on open markets. 'Corporate capitalism' is all about chaining these federated structures together into elaborate formations, using the outputs of one as the inputs to another: oil from BP's new DeepFuel subsidiary ends up as an input to plastics production by Dow Chemical, which can be combined with steel from ArcelorMittal to produce specialist extrusion machines for Nestlé's confectionery manufacturing.

These corporate-to-corporate networks form the core of the global economy. They ping-pong the components of half-finished products via transnational logistics networks until they eventually get completed and find their way via wholesalers onto the shelves of a local deli.

That's when I walk in and complete the chain by exchanging money for a milk chocolate bar (to give me energy to carry on labouring). When I buy that chocolate, I might hand money over to a person who looks like me, which may lead me to believe that we are equals participating in a mutual exchange on a market. But what I don't see is the institutional infrastructures that lay the foundations for this small act: the contract law, military force, property rights, expansive corporate supply chains and vast systems of global trade finance. I do not see that every purchase in the store is partially completing a multi-tiered financial circuit that was opened possibly decades (even centuries) ago.

The problem of partial vision

It's hard to see the interlinking elements of corporate capitalism from our street-level vantage point. We find ourselves in the position of the blindfolded sages feeling an elephant, believing the tail to be a rope, and the leg a tree trunk. Our media reinforces this compartmentalised view, speaking of 'the consumer' – a person who decides what to purchase – as a separate being from 'the employee', the same person who receives money to walk into work the next day; and separate again from 'the saver', the same person who decides to hand over control of their money to a financial institution (scanning the system for opportunities to charge up some new corporate circuit).

The consumer, the employee and the saver are the same person in the same monetary system. One day money is coming to you and on another you are sending it away, either for goods or for financial contracts, from where it is refracted out in different directions again. Financial institutions will not limit themselves to power-charging corporations, and will insert themselves anywhere in the system where there is money movement. For example, they can offer consumer financing, encouraging people to go into debt to buy products output by the corporations. They can finance wannabe workers – what is student finance but future workers going into debt in order to get into a position to be hired by a company? And they can finance mortgages for workers just free of student debt but desperate to buy an apartment to get some stability in this insecure world.

That same desire for safety can be exploited through get-rich-quick speculative investments. When I left my role as a derivatives broker, I consulted for a few months at a spread-betting company, which borrowed money from a major investment bank to lend to individual 'day traders' who wanted to escape the grind by betting in financial markets. Yes, a big financier was financing a medium financier that financed small financiers. Financing financiers is big business. You can lend to shareholders to buy shares, just as you can lend to creditors to buy bonds. It gets very fractal.

Every one of these financial contracts can then be used as raw material to build more complex contracts. We have already seen how a share in BP is a claim upon the income of 1,100 subsidiary companies, but a unit in a broader fund that contains that share might derive income from tens of thousands of subsidiary companies, while a unit in a fund-of-funds might derive income from hundreds of thousands of them. Financial institutions can pool together any existing contracts they have built – like consumer loans, student loans or mortgages – into instruments like the collateralised debt obligations and mortgage-backed securities that set off the 2008 financial crisis. These structures were built from hundreds of thousands of underlying financial instruments. And that's all before we get to derivatives – the sector I used to work in – which are bets on these different chains of contracts. An equity index swaption, for example, is a bet on a bet on the income of hundreds of corporations with hundreds of thousands of subsidiaries.

The banker having meetings about equity index swaptions is in a realm of high abstraction at the top of the economic hierarchy; people like this are more likely to be packaging thousands of mortgages into colossal bundles to resell to the Norwegian sovereign wealth fund than talking to an anxious young couple looking to buy a new home. The highest reaches of skyscrapers are occupied by very few people, but they operate at the largest scale.

The tall towers nevertheless rely upon a system of ground-level touchpoints to funnel people in. This includes the bank branches, independent financial advisers, retail stockbroker and spreadbetting companies, all of which hire customer service staff. For a corporation, however, it feels costly to provide all this ground-level human service to millions. Those people are needed collectively, but individually each one is an annoyance. This is why financial institutions look to automate their interactions with individuals. Better to provide people with a standardised digital app that allows for self-service; now they can funnel themselves into the financial mega-core, with no need for hand-holding.

But the desire to automate goes further still. It extends *into* the skyscrapers. Financial institutions are, in many ways, constrained by their unreliable employees with their unpredictable needs, feuds and dreams, sitting with their pants down on the toilet after a night of drunken revelry. Why not get rid of the stalls altogether, and instead train machines to design and package up all those financial contracts? Automate the nerve centres. That is why their towers host Level 39.

There is a small glitch standing in the way of all of this, however. It is called *cash*.

The glitch

It is crucial to split corporate capitalism into a core and a periphery. The core I have already described – mega-players like BP operating in conjunction with the financial sector – and the core players use the digital payments system run by the commercial banking sector. When BP orders a contractor to lay hundreds of kilometres worth of pipeline cement, it does not pay them in cash. It uses bulk digital bank transfers, which manifest in their contractor's bank account, from which point that contractor can ping digital bank transfers to its contractors. The chain of impulses originating in the towers of high finance crackle through a network of bank accounts, but that crackle hits a less conductive medium when it reaches the peripheries. On the peripheries of corporate capitalism are all the individual workers who do all the dirty work, and they are the ones who historically use physical cash.

Cash congregating in the vast peripheral outskirts is like an itch that the banking and corporate sector wants to scratch. Cash finds a home in the scruffy wallet of a mine labourer, or in the bra of a Xhosa grandmother buying one box of Nestlé baby formula from a tiny rural store in South Africa. Financial corporations dream of a 'cashless society' in which even these tiniest of nodes in the capitalist market will be tethered to their accounts, inserting the banking sector into every pixel of the economic picture. To entrench itself fully in our lives, and stick to our bodies, corporate capitalism needs the money to be digital, and for the ground-level touchpoints to be replaced with standardised apps, hosted in smartphones that follow us wherever we go. This is referred to as 'the future of money'.

2

The War on Cash

The scene established in the opening pages of a basic economics textbook is reminiscent of *Romeo and Juliet*. Two families – the Buyers and the Sellers – stand in opposition. The first possesses money, while the second possesses goods or services. They engage in a tense stand-off, during which the Buyers patrol back and forth along the Demand Curve – presenting the terms on which they will buy – while the Sellers stake out the Supply Curve, presenting terms on which they will sell. Where these curves intersect, Buyer meets Seller and hands over a certain amount of money for a certain quantity of goods. This point of peacemaking marks the Market Price.

Sitting on a British Airways flight, with cash in my wallet, I represent the Buyers. The cabin crew are acting for the Sellers, with a catalogue of goods for sale. The stage is set for the exchange. I make my move by requesting an instant coffee. The flight attendant disappears to prepare the beverage, before returning with it on a tray. It's my turn again. I pull out a £5 note.

'Sorry, sir – we only take cards.' The romance grinds to a halt.

With a startled look I say, 'I don't have a card.'

The attendant looks shocked and then embarrassed.

'I guess you'll have to take this back,' I say slowly. She gives me a pained expression, apologises and withdraws with the coffee. The passengers around me look awkward, too. What a forlorn deadbeat, they think: no bank card – can't even get buy himself a hot drink.

I do, in fact, have a debit card, but I believe in the process set out by economics textbooks. I'm a buyer with money and you're a seller with goods, so do a deal with me. Why must we route this relationship through Visa and the banking sector?

I return from the toilet to discover that the man next to me has volunteered to buy me the coffee using his card. The attendant is beaming at his kindness and the guy is giving me a thumbs-up. The passenger audience is grinning joyfully. I am outwardly gracious, but thinking to myself, 'Come on, I was trying to make a point!' This is not the first time my pedantic protest against British Airways' anti-cash policy has been thwarted by a card-wielding philanthropist.

The moral of the story, however, is that while markets are traditionally conceptualised by economists as involving *two* groups, the world we are moving towards introduces a *third* into every transaction – the Money-Passers, a conglomeration of banks and payments intermediaries like Visa and Mastercard. In this world Romeo and Juliet cannot kiss until a priest stands between them, joining their hands to bless the union. Unlike actual priests, though, they do this from a great distance, operating via a system of distant datacentres ('the cloud') with which we remotely interact. Their digital financial clouds are spreading like an intrusive mist over us all.

Mist

'Mist' is a useful counterpart to the cloud metaphor, because it conveys the ground-level experience of being engulfed by the lower reaches of a cloud. In the same way that a light mist is quasi-invisible, people often do not see the way digital intermediaries stand between them and others; the intermediation happens so subtly and rapidly that it seems to work just like magic.

But this magic can fail. In 2016 I was invited to speak on a panel at a conference called Reinventing Money at the Dutch University of Delft. I was jet-lagged and exhausted, and did not want to be on stage in that state, so fifteen minutes before my talk I went on a search for Coca-Cola. I found a vending machine, but it wouldn't take my cash, and came with a small digital interface built by the Dutch company Payter that said 'Contactless payment only'. I took out my card with irritation, and tapped it but – despite the fact that there was money in my bank account – the Payter device just beeped in protest: 'Card invalid.' I looked at the expiry date on my card – it was definitely still current.

Economics textbooks imagine a free market as one in which rational individuals enter into monetary exchange with each other for their mutual benefit. But here I am, a tired individual rationally seeking sugar, facing soft drinks stacked on a shelf controlled by a vending machine acting on behalf of a seller. The 'market' is here, and this machine is programmed to abide by a simple contract that says, 'If you give money to my boss, I will give you a Coke.' I have the money to do it, so this is a *market failure*. I was blocked from engaging in free trade.

Old vending machines had a little slot for coins that allowed anybody to turn their income into sustenance, even if they were homeless and without a bank account. But my Dutch vending machine was really *two* machines: its main body belongs to the seller, but to get a Coke from there I first must send a request to a complex of payments gatekeepers – including Visa and a series of banks – via the Payter card reader that is posted there to work for them. If one of those gatekeepers does not want to do business with me, I cannot do business with the seller. The payments gatekeepers can disrupt the core capitalist ritual – the transfer of money for the transfer of goods.

The Payter device also allowed no direct complaint, exuding mechanical indifference, accountable only to bosses far away. So a market transaction between a buyer and a seller was jammed by an unaccountable, incompetent and indifferent money-passer. In my Romeo and Juliet example, this is like Friar Laurence simply not turning up to the ceremony.

Alternatively, imagine if the priest fainted in the middle of the ceremony. This is akin to what happens in a systems failure: during a ten-hour outage in Visa's European systems in 2018 – caused by a failure in its primary data centre – 5.2 million payments attempts were blocked, leaving people who had become dependent upon card payments stranded and searching for ATMs (which, as we will see later, are increasingly being shut down, making them ever-harder to find). The same would happen if our electricity supply went down.

Or imagine the priest was attacked on the way to the ceremony. Digital payment opens people up to personal hacking from distant criminals using malware like Dridex (which uses Microsoft Word documents to infiltrate computers and steal bank details), and also exposes them to general cyberattacks on central systems, and mass data breaches. In February 2016 hackers used the SWIFT global payments network to withdraw roughly $1 billion from the Bangladeshi Central Bank's account at the US Federal Reserve, one of the most secure financial institutions in the world. Imagine what they could do to an ordinary account.

When talking about a market driven by supply (sellers) and demand (buyers), economists do not mention payments gatekeepers, which means they picture their protagonists using *cash,* a form of money that is first issued by states and that later percolates organically between people to catalyse markets. But the new family of digital 'money-passers' is made up of private actors seeking profit, and to use their services the seller and buyer must *buy them* from the money-passer who *sells them* for . . . well, money. This creates a strange circularity in the economic equation, because in addition to the original supply and demand, we must add supply and demand for the resolution of supply and demand.

If the money-passers can collectively entrench themselves in our payments, a slice of every single interaction in a market system will go to them, and this new age of money-passers standing in-between will be one of the most crucial changes market systems have seen in centuries. The ability to chaperone payments not only entrenches the overall power of the banking sector, but also enables three further things. First, transaction surveillance: the money-passers can monitor your transactions to collect sensitive data about your daily life. Secondly, transaction censorship: they can block transactions they do not like, and, because you do not directly hold the money, can freeze and expropriate it. Thirdly, mass automation leads to further corporate power: remote digital corporations require remote digital money.

Digital money underpinned by the banking sector is laying the foundations for the next stage of both US-dominated surveillance capitalism and its Chinese counterpart (which has higher state involvement but seeks the same outward expansion of its digital tech giants). The digital payments industry, however, makes sure never to highlight the dangers of this. Rather, to make incursions into face-to-face commerce, it showcases the surface-level 'feel' of digital payments – their slickness or apparent convenience – rather than drawing attention to the deeper structures that underpin them.

Superficial advantages

This diversionary strategy has certainly worked: try asking some-body about their views on cash vs. digital payment and they quickly veer towards describing this surface experience, giving their take on which is faster, more familiar, easier to use, more culturally attract-ive or safer. They may express some concern about the psychological effects of digital payment – for example, they may feel they spend more with their card because it seems 'less real'.

But even in this realm of surface experience, survey data shows that there remains significant divergence in public opinion on which payment method is actually easier to use. Of course, digital payment is easiest for paying people who are not in front of you – in, say, online transactions – but why insert datacentres into face-to-face street transactions? For hundreds of years people have found it easy to hand over cash, and the notion of it being 'hard to use' is laughable.

Nevertheless, digital payments promoters are likely to talk up the inconvenience of visiting an ATM, or carrying cash. There are countless adverts from digital payments companies the world over that say as much. Indirectly holding digital money is, they insist, safer than directly holding cash (provided we ignore the minor perils of cybercrime and bank failure). In 2020, these promoters took advantage of the Covid-19 pandemic to spread the idea that cash is a disease vector, despite contradictory evidence from the World Health Organisation and the German health research agency the Robert Koch Institute. Indeed, further central bank research showed that the PIN pads associated with digital payment pose a greater risk.

But even prior to Covid-19, these and other arguments were made by the Better Than Cash Alliance, a New York-based global partnership of governments, corporates and international organisations that wants to accelerate the adoption of digital payments in poorer countries. The Alliance makes some compelling points: for example, in remote areas it may be hard to establish ATMs and bank branches, and it may indeed be easier in these scenarios to keep the money in remote bank datacentres and equip people with mobile phones to control it. This digital payment is presented as being potentially more efficient for merchants, who bear various costs for handling cash, and is also tied to the concept of 'banking the unbanked' – setting up poorer people with bank accounts where they can then convert their physical money into digital money, thereby diversifying the forms of money they hold to reduce risk.

The stories told by the Alliance play into a broader story we are told, which is that the move away from cash towards digital payment systems is 'natural progress' – an organic evolution from the bottom up as a result of changing customer preferences. But what are the actual trends?

In 2019 I was invited to a gathering of central bank researchers who study cash usage. One by one they presented the trends from their home regions. The Deutsche Bundesbank went first. 'There has been an impressive increase in demand for Euro banknotes.' Central Bank of Hungary went next. 'Cash demand has been steadily increasing since 2008.' The Swiss followed. 'Growing demand for cash.' Japan. 'Cash usage remains strong.' Canada. 'Cash use is not declining.' US Federal Reserve. 'Cash continues to grow.'

In fact, the only two countries seeing absolute declines in cash demand were Norway and Sweden. This is well known in central banking circles. For example, in November 2017 the San Francisco Fed published an article entitled 'Reports on the End of Cash are Greatly Exaggerated'.

Cash usage statistics, however, are deceptive, because cash has more than one use. The researchers distinguish between *transactional usage* – the use of cash for day-to-day transactions – and *hoarding*, the storage of cash under the proverbial mattress. So here's the tricky bit. Overall cash usage has grown, but cash usage for *transactions* has, in relative terms, declined in almost all countries, sometimes drastically. People still want cash but – often – as a way of keeping their money outside the banking sector. The most recent example of the dual character of cash usage has been the Covid-19 response: central banks recorded a large increase in cash withdrawn from ATMs – because in the midst of a crisis people are scared that the banking sector will collapse – but also recorded a further reduction in transactional usage of cash.

Anti-cash crusaders may try to claim that cash hoarders are likely to be criminals but – according to the central bank researchers – a cash hoarder may be a person who does not trust the banking sector, including those who don't want their wealth locked up in banks during a banking crisis. The Federal Reserve sees huge increases in cash demand prior to hurricanes, because people do not want digital money when the power goes down. They want *offline* money. As the saying goes, 'Cash doesn't crash', and that refers both to the failure of electrical or communication systems, and the failure of banking systems.

The war on cash

My grandparents were not criminals, but they seemed entirely comfortable with cash, and never spoke of it as inconvenient. In many poorer countries, cash still dominates on all fronts, including everyday transactions. Even in the USA, prior to the pandemic it accounted for over 50 per cent of transactions under $10 and 30 per cent of overall payments volume. Many people seem stubbornly to defy the notion that cash is dead, but why do we see so many media stories on the imminent 'death of cash'?

Well, I know one thing. There are institutions that would very much like to see the death of cash. The Better Than Cash Alliance is easy to pick on in this regard. It looks like a front for the interests of the digital payments industry because, while it officially operates under the UN Capital Development Fund, it receives funding from the likes of Citigroup, Mastercard and Visa. The Alliance promotes digital bank payments in poorer countries while discouraging the use of physical state cash, and it just so happens that Visa and Mastercard have not only the same goal, but also the technology to achieve it. On a call I made to one of the Alliance directors, he downplayed this story, presenting his organisation as a small research institute working for financial inclusion, and yet it is clear that some very large companies find the mission of this little advocacy group very useful, and are happy to partner.

It would not be the first time such partnerships have formed. No Cash Day® was created in 2011 by Cashless Way – founded by corporate public relations specialist Geronimo Emili – to cast cash as perishable, unsafe, unhygienic and expensive to produce. That the day has a trademark symbol suggests it is not a grassroots initiative, and a glance at its partners list confirms this: a cohort of banks, payment companies and government agencies like Poland's Ministry of Finance and the European Parliament (which sponsored it in 2016). Emili – who also founded the website WarOnCash. org – produced a 'Manifesto for Cashlessness', and partnered with Money 2020, one of the world's biggest fintech conferences, to evangelise it.

The Penny for London charity campaign was another of these industry-led initiatives. Set up in 2014, it aimed to encourage the use of contactless payments cards on Transport for London's train system by getting people to auto-donate one penny to underprivileged children when they did so. It was chaired by Paulette Rowe, who went on to head up Facebook's global payments division, but who was then managing director at Barclaycard, the company responsible for running TfL's contactless payments infrastructure. The Penny for London directors included former CEO of Barclays Bob Diamond and hedge fund mogul Lord Stanley Fink. Hosted by the then Mayor Boris Johnson within his Mayor's Fund, the campaign included trustees from Santander Bank, Goldman Sachs, and Promontory Financial Group: here, an altruistic project is composed of a cosy group of financial insiders whose charitable efforts also happen to promote digital payments.

The examples above are just the tip of the iceberg of a lobbying and influencing infrastructure set up to wage a covert Cold War against cash. A covert Cold War sounds like a conspiracy of shady figures hosting back-room meetings. But the war on cash is hiding in plain sight. Let's for a moment, though, imagine that it were a 'conspiracy'. Who would be the 'conspirators' and what would they stand to gain?

Conspirator 1: The banking sector

Eliminating cash would lock people into full dependence on the banking sector for all payments, which means the banking industry – in general – has much to gain. Dip into the insider world of banking – read industry magazines or attend banking conferences – and you will find widespread disavowal of cash. For example, in June 2019 the Bank of America CEO Brian Moynihan declared that 'we want a cashless society', noting that his firm has 'more to gain than anybody' from a move to digital transactions.

While banks do get into industry associations to advance a common front, their more immediate concern is how to increase their individual profits. This typically entails a combination of cutting costs and boosting revenues. In terms of the former, banks see their cash and ATM operations as a cost, an irritatingly unprofitable thing they have to run in order to allow people to 'take their money out of the bank'. It would be convenient if customers slowly forgot that they should expect this right to exit, or found themselves slowly nudged out of that option. Banks try all manner of tactics to discourage cash usage, and feel increasingly entitled to penalise people for it.

On the other hand, banks have an incentive to encourage digital payment. Two major revenue sources for high-street banks are interest and fees. Credit cards give them both, and debit cards give them fee income. Their digital payments divisions are profit centres creating positive revenue. Their annual reports confirm these views, touting how they are boosting their digital payments division while cutting their heavier or costlier physical branch and ATM networks. Not only can they make greater profit by shifting people to digital channels, where they can be dealt with remotely using algorithms and customer service bots, but digital payments also create *data* about customer behaviour. Banks can use this to build customer profiles that will help them predict the behaviour of account holders and cross-sell products to them.

Conspirator 2: The payments companies

For companies like Visa and Mastercard the issue is straightforward. They make fees from facilitating money transfers between bank accounts, and see cash as competition to be eliminated. Excited by the untapped potential of cash-heavy poorer economies, they parade around the 'financial inclusion' scene proclaiming humanitarian ideals. Unlike banks, who are in general more diplomatic in their anti-cash stance, Visa's annual report is full of unabashed declarations of war against it. With zeal they talk about 'freeing' people from cash, like self-righteous crusaders freeing heathens.

They have two primary fronts in this crusade, and two primary tactics. The front they control is Internet e-commerce, which, as mentioned, is a natural home for digital payments. Payments companies ally themselves with online firms in their push to get people off the high street. The second, and far more difficult, front is street-level commerce. Cash is strongest in situations of physical proximity, so payments companies use a plethora of apps, cards and 'point-of-sale' (POS) devices to battle cash directly on its home turf.

The first battle tactic is to extol the virtues of digital payment, and to sign people up to promote it on their behalf. In India, for example, Visa has run the 'cashless man of India' campaign, and another under the banner that 'Kindness is Cashless'. The second tactic is to demonise cash. For example, in 2016 Visa UK launched its 'Cashfree and Proud' advertising campaign, with the company noting in the background that the 'campaign is the latest step of Visa UK's long-term strategy to make cash "peculiar" by 2020'. In many ways, it has succeeded. The campaign was rolled out across London billboards, radio stations and TV, and by 2019 the psychological balance in the city looked to have shifted towards digital payments. A rash of bougie hipster shops that refused to accept cash sprung up like a physical meme, helped along by incentives Visa offers (in the US this has included, for example, the company's Cashless Challenge competition, in which it handed out $10,000 prizes to small trendy businesses that 'go cashless').

In 2020 this demonisation was further accelerated when big retailers took a lead from payments companies to erroneously associate cash with Covid. For example, the major sports retailer Decathlon placed large signs at the entrance to its London megastore, reading, 'For your protection we are only accepting card payments.' This despite the fact that the Bank of England released a scientific report noting that card machines, trolley handles, goods on open shelves and the screens of self-checkout counters – all of which Decathlon has – pose a far greater risk of spreading the virus than cash does.

By the time this book is released cash may indeed seem very peculiar in London, but fifty years ago cash was seen as entirely legitimate. Fifty years ago, however, Visa was only a young company (with a different name). Over the years it, and others, have managed to install a level of moral panic about cash. The industry has consistently cast card payments as being safer, cleaner and higher status than cash, thereby slowly associating the latter with crime, disease and low status. Payments companies even spread ideas about cash as environmentally unsustainable, as if digital payment-fuelled Internet commerce has not led to massive increases in energy-intensive logistics and consumption.

Conspirator 3: The financial technology industry

The fintech industry specialises in automating broader financial services (such as automating the decision process for who gets loans), but almost all fintechs rely upon the underlying infrastructure provided by banks and payments companies. They have a natural alliance, because to automate finance in general you need to automate payment systems in particular. Put simply, digital finance does not work with non-digital payments, and fintech developers see cash as a bug standing in the way of their financial automation. The last time I visited Level 39 – the big fintech hub in London – they refused to take cash at the bar. When I asked why, they looked shocked. Fintech depends on a move away from cash. Surely it was obvious they would not accept it here!

They are not the only tech companies who depend on this, though – in 2018, Amazon lobbied against legislation requiring shops to accept cash in cities like Philadelphia. Cash does not gel with automation, which means Amazon – which is pioneering a human-free automated system the likes of which the world has never seen – does not gel with cash. This is a major theme we shall return to later.

Conspirator 4: States and central banks

In anti-state libertarian circles, the drive against cash is presented as being orchestrated by a Big Brother state that wishes to watch transactions to gain more control. Every digital bank payment is recorded in a database, leaving a clear data trail. State bodies that may desire such data include the tax authorities – to watch for tax evasion – and the security authorities, to watch for money launderers, terrorist financiers or, alternatively, political dissidents (pro-democracy campaigners, minority rights groups, environmental activists and so on). Then there is the central bank, which may wish for greater general surveillance over a country's economic activity and greater monetary policy control.

Libertarians are not wrong to have these suspicions, because there is plenty of evidence of overt state action against cash. Twelve EU member states have implemented 'cash thresholds' to prevent the use of cash over a certain amount (for example, €1,000 in Portugal) under the justification of preventing terrorist financing and money laundering. Others with thresholds include Uruguay, Mexico and Jamaica, and India and Russia have proposed them too. These thresholds are designed to ratchet down over time, with the cap slowly being reduced to wean people off cash purchases. For example, Greece started with a cash threshold of €1,500 and then moved it down to €500, and recently proposed to ratchet it down further to €300.

Germany, notably, has vigorously resisted cash thresholds. This is interesting, because anti-cash pundits present the demand for cash as being driven by underworld crime and corruption, and yet

Germany ranks ninth out of 180 countries on the 2020 Corruption Perceptions Index, meaning it is perceived as highly trustworthy. The paternalistic message in places like Italy and Greece is that people who want cash are dodgy tax evaders, while Germans who want the same thing are presented as privacy-aware or prudently keeping their savings under a mattress.

No state has been bold enough to pass a law banning cash, but many have created national strategies to transition away from it. France has a national 'cashless payments strategy', and the Greek state (under huge pressure from creditors to repay onerous debt incurred through the opportunistic lending of German and French banks) has taken to a hostile anti-cash rhetoric to eliminate petty tax evasion. Other openly anti-cash governments include Nigeria and Hungary.

To complement the anti-cash rhetoric, these states praise the digital payments industry and encourage adoption of digital banking. Their support takes different forms, from state-funded digital innovation hubs through to paying welfare recipients via digital payments platforms, and it is often coupled with a refusal to accept state cash for state services.

Countries like Sweden have also made use of 'demonetisations' – invalidating older notes to force people to change them – an action that creates inconvenience and uncertainty around cash. The most dramatic use of 'demonetisation', though, was that undertaken by India in 2016, when the Modi government sent a severe shock into the cash system by illegalising certain notes overnight and forcing people to turn them in within a very short period. This action caused profound disruption in the lives of poorer people who depend on cash, while sending a powerful negative message about it – officially presenting it as the 'black money' of corruption. The Modi government used the patriotic press to push anti-cash sentiment, before spinning the story into a tale of aspirational digital modernity: it preached about a bright, efficient and convenient cashless future that people would arrive at whether they wanted to or not.

Not surprisingly, this gave the Indian digital payments industry a massive boost, and the industry reciprocated with sycophantic front-page adverts in praise of Prime Minister Modi. The Indian digital payments giant Paytm put a full-page advertisement on the front of the *Times of India* and the *Hindustan Times* saying: 'Paytm congratulates Honorable Prime Minister Sh. Narendra Modi on taking the boldest decision in the financial history of Independent India! Join the revolution!'

It was not only Indian corporations sidling up to Modi. In its 2017 annual report, Visa noted that 'During the year, we worked closely with the Indian government to support its demonetisation efforts,' a process that led to the *doubling* of merchant acceptance of Visa in the country.

The Indian government has also been vigorously pushing through the world's largest biometrics programme – Aadhar – which is justified in terms of financial modernisation: digital payment accounts require people to verify their identity, and fingerprints and iris scans are a way for illiterate people to do just that. It was initially pitched as a voluntary complement to other ID systems, but over time Aadhar has increasingly become a requirement to access basic government services, and also prompted the largest supreme court privacy dispute in modern Indian history.

The Indian state has legitimate reasons to push economic development programmes, but there is no denying that these programmes simultaneously promote the commercial interests of the digital financial sector. The country is not alone. These alliances of states and financial institutions are found the world over and are enmeshed with major international development institutions like the Bill & Melinda Gates Foundation. The latter also funds the

Better Than Cash Alliance, alongside USAID, the US government development agency. Once we begin to look for them, the anti-cash stances of the various 'conspirators' outlined in the sections above are easy to spot. Their top-down initiatives, from anti-cash advertising campaigns to pro-digital policy recommendations, should immediately cast doubt over the extent to which a move away from cash is truly a bottom-up phenomenon.

Aesop's paradox

The Better Than Cash Alliance invites Bill Gates to promote the digital payments industry in front of delegations of policymakers, providing countries (some with shaky democratic credentials) with respectable cover to promote their own anti-cash programmes (Malaysia, for example, has reputedly explored demonetisation). As this saga unfolds, I cannot help but be reminded of the ancient Greek fable of the horse and the stag, as told by Aesop:

> The horse has a quarrel with the stag, and so seeks the help of a hunter to kill the stag. The hunter agrees to collaborate, but suggests that to do so he must first put a saddle on the horse's back and a bit in its mouth to co-ordinate the attack. The horse agrees. They kill the stag, and then the hunter – with a glint in his eye – says, 'I think I may just leave this saddle on you.'

Partnering with a powerful party in order to combat an imagined enemy can leave you inextricably bound to your new 'partner', and in a subservient position. And so it is with states who uncritically promote the digital payments industry (i.e., the banking industry) as a weapon to defeat the scourge of tax evasion or criminal transactions (or, now, pandemics).

But there is a crucial difference between impending cashlessness and the original story. In Aesop's fable, the hunter and horse kill the stag. But the digital payments alliance does not kill crime, corruption and tax avoidance. If anything, financial cybercrime is flourishing, and the very banks that underpin the digital payments infrastructure are the ones who also offer services to corporations to facilitate large-scale offshore tax avoidance.

That said, local and state governments are not monolithic entities. While they may have control agendas (watching people), and crony tendencies (supporting private corporations), in democratic countries they also experience a counter-pull from the public in the form of mandates to look after people via consumer protection laws, competition laws and principles of human rights, privacy, and inclusion. For example, a number of American city governments have gone against the industry grain and moved to protect cash on inclusion grounds, by requiring shop-owners to accept it so as not to exclude those without bank accounts.* Public bodies (particularly local or smaller ones) can be surprisingly concerned about public opinion, and unlike corporations – which have narrow agendas – they often have competing agencies with conflicting mandates. For example, I met an Israeli central banker who expressed concern that the Israeli tax authorities were urging people not to use cash, noting that many in the central bank felt that this was irresponsible public messaging.

In the 1980s it became fashionable to express disbelief that political figures could serve anyone but themselves, but – at least in my experience – many civil servants are still more likely to see themselves as holding a responsibility towards the public than a digital

---

* These cities include Philadelphia, San Francisco and New York. The states of New Jersey and Massachusetts have also passed state-wide legislation to prevent cashless stores

payments company. On a trip to Kenya, for instance, I met deputy governors from four East African central banks – the East African digital payments community is booming, but the central bankers are among the only ones raising privacy and consumer protection concerns. The German Bundesbank is notable for its support of cash, and I have had private conversations with Bundesbank employees who talk seriously about having a public responsibility to protect it.

Nevertheless, various insiders are urging central banks to turn their backs on cash. Cash roams the world with us 'offline'. Somebody cannot push a button and make the note in my hand disintegrate. They can, however, do that with digital money, because it is recorded in a datacentre under the control of a financial institution that can alter it. A number of macro-economists are excited about this latter property, because it opens up the potential for 'negative interest rates' – the ability to get banks to erode people's digital deposits, which, if done during a recession, might hypothetically inspire people to spend rather than hoard money. Along with researchers at the IMF, the Harvard professor Kenneth Rogoff has been at the forefront of attacking cash on these grounds, arguing that it stands in the way of this hidden monetary policy tool.

Despite this, many central bankers officially describe themselves as 'neutral' on cash, neither overtly supporting it nor discouraging it. The problem, though, is that being neutral on an uneven playing field is roughly the same as supporting the most powerful party. Imagine an older brother bullying his younger sister while the parents insist on being 'neutral', neither supporting nor hindering either party. That is effectively taking the brother's side.

Cash has very few institutions fighting its corner. There are no venture capitalists who make profit from it, but there are a *lot* of venture capitalists funding private payments companies who have a big commercial interest in flooding the media with negative stories about cash. The central banks *issue* cash, but appear reluctant to promote it in an official or unified manner, lest they seem biased. In their neutrality they are letting the digital payments industry – run by their banking members – take over. Critical observers might see the official neutrality of central banks as a covert way to slowly get rid of the cash system, with their lack of action an action in itself. Authorities have not stopped banks shutting down ATMs, and often talk about it as a private sector matter rather than a matter of public concern. Their position in many countries has largely been to step aside and let the hounds of digital payment tear into the cash system.

But while the commercial players approach the war on cash like crusaders, the state support for them is not without angst. States find themselves getting edgy in their alliance with the digital payments industry as security advisers raise growing concerns about cyber attacks, payments infrastructure failures or even terrorist attacks on digital payments systems that could bring economies to a halt. Then there are questions of financial stability. What if there is a banking crisis in the context of cashlessness – how will people exit the banking system if there are no ATMs? After years of letting cash get trampled in their economies, it is now the Swedish and Dutch central banks voicing reservations about its disappearance. In 2018 the Swedish authorities went so far as to release a pamphlet called 'If Crisis or War Comes', in which they suggest citizens may wish to keep some cash on hand for emergencies, such as a showdown with Russia.

But alas, state authorities – like Aesop's horse – may have got themselves too far into the alliance to back down. Understanding the politics of this requires us to take a journey into the core of the monetary system. What is the nature of cash, and how does it differ from digital money? And, most importantly, what is money, and how does it work?

3

The Giant in the Mountain

On the train to work each day my osteopath John is surrounded by Londoners with poor posture, staring down at their phones in a way that – unbeknown to them – is causing long-term damage to their neck vertebrae. He'd try to warn them, but they've already got so many stresses, and it is only when the harm is done, and the excruciating pain hits, that they will turn up at his clinic.

That is why I'm here on his couch. I spend a great deal of time hunched with poor posture in front of my computer, trying to alert people to the dangers of financial institutions they are too busy to think much about. In the short-term bustle of our lives, we are seldom aware of the damages that are slowly accruing, whether to our vertebrae, or to our freedom.

John finishes the session with me, and holds out a device. 'Cash or card?' he asks. He doesn't realise how much this question feeds into my stress levels.

'Pay by card' is an odd phrase, because what we really mean is that we are paying by *digital bank money transfer*, which is initiated – but not completed – by presenting a card. Asking 'Cash or card?' is like asking, 'Will you travel by bicycle or car-key?' In the Netherlands they say 'pay by PIN', which is like saying 'travel by driving licence'. To all appearances, using a card (or another payment initiation device like a phone) makes digital money fly through the air. But under the hood we find a convoluted transnational payments circuitry, tied together by institutions you cannot see, but who can see you.

So what is the difference between payment by cash and digital bank money transfer? The former takes place *here and now* using state-issued money, whereas the latter takes place *elsewhere and later* using bank-issued money, which is an entirely different form of money. This distinction reflects the fact that we live under a symbiotic dual monetary system, consisting of two separate systems fused together. When John asks me 'cash or card', he is asking whether I wish to use the primary core system run by the state, or the secondary system wrapped around that, run by the banking sector (there is also a tertiary system wrapped around the secondary system, run by the likes of PayPal). In truth these primary and secondary systems operate as an inseparable complex, but to understand how they interlock we must tease them apart. The rest of this chapter is dedicated to describing the primary core, and the chapter after will fill in the rest.

Money users vs. money issuers

The incantation 'Abracadabra' is reputedly derived from the Aramaic for 'I will create as I speak'. Speaking is not only a means to describe things, request things or order people around. It can be a means to *create things*, and a well-known example of this 'conjuring through words' is *the promise*. A promise is a 'thing' created by the act of uttering it. You *give* a promise to somebody who receives it. This can be very strong, and it can even induce the person to give you something in return.

Notice that a promise is two-sided. A *promiser* brings it to life by speaking it to a *promisee*, or, alternatively, by writing it down and handing it to them. This gives us the key to grasping modern money systems. We grow up with a one-sided view of money as a self-contained commodity-like object floating around in the world. To grow into mature monetary thinking, money must be seen as *two-sided*. One side is the money issuer, and the other is us – the money user.

The primary money issuer in our society is the state, which issues money via the treasury and central bank (the main exception to this being the Eurozone, in which a 'meta-state' central bank issues money). Many central banks present themselves as being 'independent' of governments, and many countries do have laws that grant the central bank a degree of immunity to act as a quasi-autonomous institution. But in the final analysis a central bank is granted its power by governments and occupies a crucial position between the government and the banking sector: it is the 'government's bank', but also the 'bank for banks' (where commercial banks gather), and therefore functions like a membership club, mediating between the state and the banking sector.

The central bank acts as an agent of the state, and jointly they issue fiat currency. To issue 'by fiat' means to issue by decree. They write money into existence, as it were. Some of it they write on paper and metal, but much of it they write on computers.

Many money users feel a degree of moral horror – or at least confusion – at the notion that money could simply be conjured into being. (This is especially the case when we're stuck in the one-sided phase of monetary thinking.) It strikes us as alien because we experience ourselves working to get money, and we have built entire cultures around that concept.

Thinking like a money user, however, is a block to understanding money. To step into two-sided monetary thinking, it is helpful to imagine yourself in the shoes of a money issuer. While this is simple, it's also counter-intuitive: you need to imagine what it must be like to *pay by promise,* writing out promises and handing them out in exchange for real things. Imagine these units emanating out from you, circulating beyond yourself, before eventually returning back. Money users are used to experiencing money as an *asset* – something they try to grab and hold onto – whereas money issuers experience it as a *liability* – something they issue outwards, and which puts them on the hook.

Much like a promise does not exist until it is granted outwards by a promiser, money only becomes money when it *moves away from* a money issuer. When it returns to the issuer it is destroyed (consider what happens when someone hands back a promise you have given them). But what is the 'promise' being issued out in the case of state fiat money?

The giant in the mountain

Imagine an ancient fantasy land of small communities who farm in separate valleys under the shadow of a large mountain. They are subsistence farmers living frugal lives, and confine themselves to their own valleys. One year, however, they are menaced by a terrifying giant, who emerges from his fortress in the high reaches of the mountain and puts everybody in the valleys below under a powerful spell. This spell requires each person to take a yearly sip of magical water from a spring within the mountain fortress, and if they fail to drink this each year, they will turn to stone. They can go about their day-to-day lives just fine, but this knowledge now sits in the back of their mind and creates an existential anxiety. In this context, they see the magical water as an *antidote.*

They only need access to this antidote once a year, though, and the giant hits upon an elegant plan. He decides he will issue tickets granting people future access to the spring, but only on condition that they first bring him tribute in the form of agricultural goods and various services. The people readily seek out these tickets – because it gives them future access to the antidote – and so the giant accumulates sheep and wine as the people strive to accumulate his tickets.

Once those tokens are in their hands, though, the people discover that they can use them – in the interim, and away from the giant's fortress – to facilitate exchange between themselves down in their villages. Because the power of the giant is spread across *all* the valleys, and the separate communities have a common requirement to get them, the tokens can move easily between the different villages. Over time they find the boundaries between villages slowly dissolving. The previously autonomous communities begin to segue into one much larger interdependent web that surrounds the mountain. Their trade begins to blossom. The more this develops, the more they become distanced from their old means of subsistence, and the more they come to rely upon specialised strangers for their survival.

After a few generations they largely forget about the giant, who now simply appears as a grumbling old man in the hills. More important to them is their expanding economic system, which now takes the form of a strong and growing network of people around the mountain, trading and specialising in a way their ancestors did not. The tickets have come to have *two* meanings: one is 'ticket to get magic water from giant', while the other is 'object that enables me to access stuff from strangers across the valley'. What was once merely a promise for a particular thing from the giant, has morphed into a much more general *economic network access token* that encompasses tens of thousands of people.

Notice that this is a better situation for the giant. Imagine he had to stage raids on the villages to pilfer potatoes and sheep. It would scare the villagers into fleeing, and disrupt their farming, which would backfire on the giant, who relies on their produce. It is more sophisticated to enchant them with a vague future threat, give them time to respond to it, and offer them a means to resolve it peacefully. They give him stuff, and as a bonus the antidote tickets are used to catalyse a broader economy, which boosts their production, which means the giant ends up getting more things. You might say it is a win-win situation. It will most likely lead to the emergence of a powerful class of capitalists in the valleys, who will grow rich on the giant's tickets (after which they may start to complain that the giant is trying to steal their tickets), and who will use those as the basis for a financial system.

Interpreting the parable

The parable is fantastical, but it conveys the idea that monetary units are like 'tickets' issued outwards (from above), and it also conveys the idea that *markets form in response*, as these tickets are subsequently transferable between previously separate communities. This is a direct inversion of the story that comes out of mainstream economics, which rests on the assumption that markets *naturally exist*, and then uses these as an explanation for why money must have come into the picture (in particular, economists consistently imagine that money spontaneously emerged to replace barter, a crude theory that has been repeatedly rejected by anthropologists). In the economics narrative, money is a *product* of markets, rather than the other way around.

While the history of different monetary systems is diverse and complex, a useful starting point is to see money as originally issued with a political agenda, after which it comes to catalyse powerful markets. Real-life examples of our parable can be found in the experience of pre-capitalist societies that did not use money – until they were colonised. Colonial officials would turn up (like the giant in the mountain) and decree that they must hand over unfamiliar tokens (issued by the colonial state) to tax officials at set points of the year, or else face punishment. How could the colonised people get hold of the tokens? Well, colonial plantation owners had them. If the people offered their labour to the plantation, they would receive the tokens required to fulfil the obligation.

Just like the giant does not want his own tickets, the colonial governments did not want their own money. They wanted to harness *labour* for the plantation owners, and rather than directly coercing it, they indirectly did so via a requirement to gather abstract tokens. In my own country of South Africa the colonial authorities used to refer to this coercive mechanism as the 'hut tax', and it was the main way to get previously tribal subsistence farmers to work in the mines owned by the local white capitalists.

The net effect of this is that previously autonomous people offered their labour to a colonial force, moved away from their traditional economies, and were incorporated into the fluid trading networks that we associate with capitalism. These monetary systems dissolve smaller communities into larger networks of people who are strangers to each other. This is what kick-starts the system, and after that it can take on a life of its own. Given that production expands when small communities of people fragment into dispersed networks of strangers that specialise and trade, the state gets far more resources via these means than if it were a mere warlord pillaging subsistence farmers.

In formerly colonised societies there are still some elders who experienced the arrival of money, but most people in advanced industrial countries have never directly experienced this. In Europe, for example, the monetary grooves in people's minds are long-established (picture Roman emperors imposing taxes on conquered

Gallic tribes). This may not be foremost in our minds when using state money – and even politicians frequently do not understand this – but political power forms part of the magnetic core of the monetary mechanism.

The fluid trading networks unleashed by monetary systems, though, feed back into states. Modern states have been partially pacified by their own populations – imagine the giant in the mountain becoming dependent upon the people he has enchanted, and forced by an increasingly strong citizen base into democratic concessions, until he becomes a mere figurehead (somewhat like modern constitutional monarchies). Nevertheless, from birth we all carry an ongoing obligation to periodically honour the state by handing back tokens it issues out. Failing to do this can lead to harsh consequences (such as being imprisoned for tax evasion or having your assets seized), which is why it is best to ignore the vague inscriptions written on state banknotes, and read them rather as 'I, the state, owe freedom from your tribute obligation, to you.' That phrase can be shortened to 'I owe you freedom,' and can be further shortened to *IOU freedom*. IOU is just a more technical way to describe a *promise*: the 'I' is the issuer. The 'O' is the promise for the thing. The 'U' is the person being promised to.

This fiat money is written into existence, but it is also spent into existence. The giant does not hand out his tickets for nothing, and states do not hand out these IOUs for nothing. They issue them in exchange for real goods and services, like road building services and agricultural goods (or, in olden times, military service – from where soldiers would distribute coins at bars and brothels, from where it would enter broader circulation). In modern societies, states use the central bank to credit the bank accounts of major companies who secure contracts from them, and from there those companies will pay workers, and so on, allowing the money to percolate outwards.

Expansion and contraction

Whenever a state spends, it is *pushing out* money, but sometimes it will accompany that with a simultaneous *pulling in*. This happens any time a state calls money back to itself, which it can do in a number of ways (including through taxation, fines, fees, and also by 'borrowing' it back). The next step on the journey to maturity in monetary thinking is to see that states are not only money issuers, but also money destroyers. Imagine the giant in the mountain pushing the tickets out but also pulling them in, a dynamic process that grows as the community surrounding the mountain expands outwards in concentric rings over time. Ancient states were like little giants surrounded by hundreds of thousands of people, but nowadays hundreds of millions of people can be held together under common money systems.

The life of a self-sufficient hunter-gatherer living off the open land 7,000 years ago would have felt rather different to how we experience ourselves nowadays. You would wake up in the morning to see a landscape unclaimed by property developers, and your imagination would be occupied with seeking out means to survive within it. We, on the other hand, wake up on parcels of land owned by people, and we have to scan our environment for opportunities to trade, because if we cannot find those, we cannot survive. This takes us back to the nervous system metaphor I developed in Chapter 1. After a few hundred years of being tied into a monetary system – through which much larger-scale webs of deeply interdependent people develop – we cannot operate without it.

Once these webs reach a certain size, the awareness of how they are held together fades away. A child first sees money as a mysterious object with a curious ability to command goods from other people, and from this perspective a government asking for that money looks roughly similar to a shopkeeper asking for it. But that is an error of perception. The belief that the state *desires* your money is a psychological glitch, akin to a squirrel believing an oak tree desires its own acorns. The state *wants you to take its money*, and has no primary interest in getting it back, other than for technical reasons of keeping the money supply taut.

For many of us this is as strange as being told for the first time that the sun does not actually rise, but rather sits suspended while the earth turns. Just as we are prone to the illusion of a sunrise, we are prone to thinking that we must give money to the state so that it can 'spend', but it is the other way around. The state 'spends money into existence' and then later calls it back out of circulation so that it can re-issue it elsewhere (or preserve the power of the money remaining in circulation). The main consideration that constrains the state is its need to not destroy – through excessive issuance – the web it depends upon. The dark art of 'monetary policy' is all about keeping that web taut enough for people within it to feel safe, and yet flexible enough that it can expand and morph.

Digital versus physical promises

The media often reports on state money issuance as 'printing money', but much state money issuance – in the first instance – is *digital,* and it is not issued to us. In other words, when spending money into existence, state representatives do not normally walk onto the street and hand out cash for goods. Rather, they do it indirectly and digitally: imagine, for example, you are a contractor hired by the central government to build a hospital. The government does not pay *you*. It pays *your bank*. It issues digital units into the account of your bank at the central bank, and leaves it up to your bank to then credit your account (a process we'll look at in the next chapter).

These digital units issued to and held by your bank are called – in monetary policy jargon – 'reserves'. They circulate in a closed system at the central bank, which is akin to a private membership club accessible only to the big commercial banks and government. Those reserves, however, can be 'materialised' into physical cash. When banks ask, a central bank can delete a promise written down on a computer (reserves) and re-write it instead on a paper bill (cash).

Physical cash is the materialised version of a state IOU, whereas digital reserves are the dematerialised version of exactly the same thing in a central bank data centre. Banks can thus interchange digital reserves for cash, and vice versa – they can hand cash back to the central bank and get it re-written on a computer instead. One of the hardest things for us ordinary money users to understand is that when those cash tokens are returned by banks to the central bank, *they are not money any more*. A crate of dollar bills sitting at the central bank is only potential money, waiting to be activated by issuance (a process that will be echoed by the destruction of digital reserves as they get 'materialised'). The power of this money does not reside in the piece of paper itself. That is just one substrate upon which a state promise can be recorded. It is all the structures of law and power within which the subsequent token circulates that will give weight to it, as well as broader network effects that we'll explore later.

Adventures of a banknote

The physical substrate, however, must be manufactured, a task that governments often outsource to private businesses. In 2017 I attended the Currency Conference in Kuala Lumpur, a forum for these secretive companies. While there I found myself seated at a dinner with a delegation from the Royal Mint, the esteemed manufacturer of British coins. The Royal Mint is a commercial enterprise and, rather than being solely loyal to the British Crown, it also acts as an international mercenary, providing minting services for over fifty other nations. It *makes* their coins, but leaves it up to them to *issue* those coins.

The Royal Mint's presence at the Currency Conference was unusual, however, because the event is primarily intended to be a global forum at which *banknote* companies pitch their services to central bankers. In many countries, the central bank is the issuer of banknotes, while the treasury is the issuer of coins. There is an academic debate about whether that makes any difference – and big debates about the extent to which central banks are truly 'independent' of government treasuries – but when central bankers want notes manufactured, they call on banknote specialists (different companies are responsible for the paper, the security features, and the inks) while the treasury contracts a mint.

Regardless of these behind-the-scenes nuances, we experience coins and notes as interchangeable state money, and both carry the same legal-tender status, being legally guaranteed to extinguish tax obligations and debts. Both must also inspire confidence by being difficult to counterfeit. Counterfeits are illegal tokens that attempt to piggyback off an existing money network by mimicking the appearance of its tokens. They can intermingle with state cash, but states see them as a parasite – the monetary web can live with them, but it is not ideal. Indeed, counterfeits have even been used as political weapons: during the American Revolution British forces deliberately counterfeited the Continental Currency of the rebel forces to sow doubt about it among its users (and thereby undermine exchange networks). In contemporary times Indian politicians accuse Pakistan's secret services of feeding counterfeits into India to engage in 'economic warfare'.

The design of cash tokens must be complex enough to make copying difficult, but not so complex that they confuse users. A banknote is like a high-security artwork printed on a very small canvas. Their designers can draw upon a whole palette of features, from intaglio engraving to produce portraits, to advanced inks with iridescent colour effects normally only found on tropical birds, to security foils that wink when tilted. The aim is to produce the visual equivalent of an intricate yet elegant haiku: each element must be carefully placed within the restrictive boundaries of what is technically possible and cost-effective to print, and many banknotes are truly beautiful pieces of work from a design perspective.

This *production* of cash tokens, however, is not the same as their *issuance*. Banknote manufacture is the production of 'potential promises'. These are delivered to the central bank, which will store them ready to be issued out. Trying to steal banknotes from a central bank is like trying to steal a promise not yet uttered. This is what happens in the 2017 Netflix series *Money Heist,* which portrays a group of renegades who lay siege to a Spanish government money-printer. The series should rather be called *Forced Money Creation*, because – as we have already discussed – the notes are not money until they leave the issuer.

Cash will normally exit the central bank when commercial banks order it in anticipation of requests from their own customers to get cash from the ATM or bank branch. Anticipating this process can be tricky: while the daily ebb and flow of demand for cash out of ATMs may be relatively predictable, there can be anomalies. As mentioned in the previous chapter, cash demand in certain regions of the USA can increase massively as a major storm approaches. People – quite rightfully – get concerned about electricity blackouts that will render digital payment systems useless. They want cash on hand.

Cash-in-transit (CIT) companies run the arterial pathways to get bulk cash from the central bank to the bank branches and ATMs on behalf of the commercial banks who order it. This opens up the more traditional way for bandits to illegally obtain state money. Growing up in South Africa, I remember scenes on local television showing gangsters with automatic rifles having shoot-outs with armoured CIT vehicles. In countries with high poverty and organised crime, the CIT industry can be risky business, and South Africa's CIT operators are often staffed by ex-soldiers, tough guys who otherwise might be hired to drive convoys of negotiators in war zones. In some countries there are other hurdles, such as geography. In Greece, banks must arrange for ATMs on tiny tourist islands to be stocked, a task that requires transporting cash on ferries.

Banks also decide what note denominations they will stock in their ATMs, a decision that affects the overall mix of notes in circulation. Giving out too many high-denomination notes causes problems for the provision of change, so UK banks like the Royal Bank of Scotland only stock big £50 notes in ATMs found in rich areas like London's Canary Wharf. The UK government also encourages them to make £5 notes available in poorer areas to solve the '£7.56 problem' – the situation in which people who have less than £10 in their account shy away from entering a bank branch to withdraw their last pounds.

Once all this background preparation is done, a banknote is ready to exit the ATM and jump into someone's wallet. From that point on it is like an emissary of the state far from home. It could spend years in transit, hitching around like a streetwise vagabond. It might even escape the country with a tourist who forgets to spend it, and be found decades later in a shoebox uncovered by their grandchildren. After generations of cash usage, many people build an emotional attachment to the symbolism of the tokens, but also to their feel. Cash is tactile, and for a lot of us there is an emotional association between physicality and certainty. This tactility is also very helpful for visually impaired people who use their fingertips to sense out quantities of money in a wallet through the shape, size and texture of different denominations (in contrast to a smart-phone screen, which feels the same no matter what).

While this physicality is an important reason why many people still favour cash, it also opens it to lines of attack, and wear and tear. The latter depends on how often cash is issued into circulation and vacuumed out again. If banknotes circulate for a long time before returning they get worn out, but if they frequently return they can be rewritten afresh. For example, a note may exit the ATM and immediately be spent at a supermarket retailer, who deposits it back into a bank, which may return it to the central bank to swap it for digital reserves. This returns the cash to its starting position, a potential promise, out of circulation, waiting to be 'activated' again (alternatively, it might be shredded and replaced with fresh notes).

The giant is not alone

This process of cash entering society and exiting again is called the 'cash cycle', and it requires the co-operation of commercial banks, who – as we shall see – are becoming increasingly unco-operative. To understand why, we need to introduce a new piece of the puzzle.

In this chapter I've presented the state as if it were the sole money issuer, but this is not the case. The state may be the *primary* money issuer, but *secondary* money issuers exist alongside it.

In the parable of the giant, I noted that the villagers can eventu-ally become powerful capitalists, and use the giant's tokens as the basis for a financial system, but I did not expand upon just how powerful that financial system can become. Indeed, it is time to leave the imagery of the giant behind, because in reality the state 'giant' has long been overtaken and has delegated its power over the monetary system to a domineering group of capitalist firms. Our monetary system today has a dual and symbiotic structure, because the state allows *commercial banks* to issue a large part of the money supply. And when I say 'commercial banks issue money', I'm not referring to the process in which they dish out state cash from the ATM. Rather, I mean that they issue out *their own digital money* to us. The ascendency of that secondary system is what they promote under the name of a 'cashless society'.

4

Digital Chips

I first saw plastic banknotes floating in a wishing well at a Transylvanian monastery. Such wells are thought to be sacred, and the ancient tradition was to drop a coin in while incanting a wish. As a beneficial side effect, the coins would infuse the water with biocidal traces of copper or silver that combatted water-borne disease, and thereby reinforced the well's perceived sanctity.

The practice of throwing banknotes into Romanian wells is, by contrast, new. It began in 1999, when the country's central bank began issuing polymer notes; unlike coins, the plastic floats, and does not infuse the water with holy trace elements. Polymer seems impervious to wear and tear, unlike older paper banknotes that have a way of ageing organically as those who hold them stuff them into wallets, pass them on to others, and send them on adventures across countries. Indeed, the wandering and gregarious nature of banknotes was captured in 1770, when Thomas Bridges published a novel called *The Adventures of a Banknote*, in which a sentient note narrates its life in Georgian Britain, as it travels the country from hand to hand.

At the time Bridges was writing, market capitalism was in full swing in Britain, and yet many people would live their entire lives without directly using a commercial bank. Private banks targeted richer merchants and aristocrats – offering them various means to settle debts and make payments to each other – while poorer workers relied upon coins and notes of various sorts. This was still the case in the early 1900s, even though banking had become more standardised by then. To this day there are still many people who do not have bank accounts or do not use them frequently. In Romania, for example, 86 per cent of people have bank accounts, but only 67 per cent use those accounts for payments, and then irregularly. Big institutions – employers – pay salaries into those bank accounts, but workers then withdraw cash and use it.

But while the spread of banking remains uneven, the trend over time has been for banks to slowly start targeting less-wealthy people and to promote mass retail banking, advertising bank accounts as a place to 'store your cash'.

In the early 1900s a bank account was literally an entry in a physical ledger book, but with the advent of computers banks have converted these paper books into computer databases. To get a little slot on that database, I have to provide proof of who I am, after which the bank gives me an *address* in the system, marked by a unique identifier called an account number. It then tethers me to this account by giving me means, such as a PIN code, to prove my association with it.

This account – apparently – is where I then 'deposit cash', and also – apparently – where the subsequent 'bank deposit' is stored. Herein, though, lies linguistic trickery and deception.

Banks are not like cloakrooms. They are like casinos

Many of us are prone to thinking of bank deposits as 'cash I put in the bank', using phrases like 'I've got cash in my account', or 'I'm going to withdraw my cash'. This language suggests *safe keeping*, as though the bank were a custodian for our money. The imagery is similar to that of a cloakroom where you might deposit your coat as you enter an event space: you hand your garment to a clerk who gives you a paper slip to recognise as much, and which gives you the right to retrieve your coat when you leave.

To believe you have 'cash in the bank' is to assume that a bank works in a similar way to a cloakroom, and that opening an account is like getting a little space where the bank stores your cash for safe-keeping, like a coat on its own shelf. Also, in this imagination, there is but one form of money – state money. It may be out of the bank, or in the bank, but banks are seen as mere intermediaries who either store it, move it around or lend it out.

But there is no such thing as 'cash I have in the bank', and my bank account does not 'store' state money. Its sole function is to record an entirely different form of money called *bank money* (or, in conjugated form, bank-money).

The quickest way to grasp this is to cast aside the cloakroom metaphor and replace it with a casino metaphor. Imagine yourself walking into a casino and handing over cash for chips that you can use inside the casino. While you're in the building, the casino is not looking after 'your cash'. It has taken ownership of it. All you own is the *chips* it has issued to you. There are thus *two* forms of money in this casino. State money, and casino-issued chips that can be redeemed for state money if you take them back to the cashier (assuming you have not lost them to other punters). The gamblers own the latter. The casino owns the former. There is a symbiotic relationship between the two forms, but they are nevertheless separate.

Normal casino chips are physical, but imagine now that a casino ran a system of *digital chips*. Imagine that, rather than handing out physical chips when you gave them cash, they opened an account for you in their computer and credited that account with digital units that can be used at the various tables. All you now own is an address on their database, with credits attributed to it.

This metaphor is a great entry point for understanding modern banking. The units we see in our bank account are just 'digital chips' *issued out* by banks, in much the same way that casinos issue me chips when I hand over cash. A casino chip is an IOU that can be redeemed for cash at some point (if the casino refused to do so, you could take them to court). Similarly, the digital chips in our bank account are IOUs written out in digital form, promising to pay out state money in the future (think of the IOU as 'I, the bank, O state money, to U'). These digital IOU chips can then be passed around.

Thus, when I open an account for the first time and see '0', it just means 'We, the bank, have not issued you any digital chips.' If I hand over £100 in cash, the bank takes ownership of it, which means the bank's stock of state money increases, but it will then issue me 100 digital chips, which will appear in my account as '100'. These chips are *promises for state money*, so while I might experience them as something I now own, the bank sees them as something that puts them on the hook. The bank has gained state-issued money, while I have gained bank-issued digital chips.

Going to the ATM is the equivalent of trying to 'exit the casino': the bank hands me state cash while retracting its digital chips, thereby nullifying the promises they previously issued to me. This sheds further light on the process described in the last chapter, whereby banks must anticipate exits and prepare for them by 'materialising' a portion of the digital reserves they hold into cash. This is the shadow side of the state cash cycle.

In an actual casino you would be able to easily distinguish between the casino-issued chip and the state-issued money you had handed in, but in the world of banking we struggle to make a distinction between bank-issued digital chips and state-issued cash. In common parlance, both are generically referred to as 'money', but there is a distinction made in technical jargon between the two: state money is called *base money* (or narrow money, high-powered money, or M0), while those bank chips are called bank-money, book money or sometimes broad money.

Perhaps the most confusing term for bank chips, though, is *bank deposits*. It is confusing because the English language has a convention by which the verb 'deposit' creates a noun of the same name. For example, a flood might *deposit* sand in a river, after which we might say there is a *sand deposit* in the river. In the case of banking, though, this convention creates havoc. We use the *verb* 'deposit' to refer to the act of 'putting in' (*I deposited cash*), and then we often erroneously use 'deposits' as a noun to refer to 'the thing put in' (*I have cash deposits in the bank*). In reality, the noun 'bank deposit' refers to the *promise issued out*, not the *thing put in*. The technically-minded reader can confirm this by perusing a bank balance sheet, where 'customer deposits' always appear on the *liability* side (where things the bank has promised to people are recorded). For the non-technical reader, just remember that 'bank deposits' are digital chips issued out by banks.

Conjuring chips 'from nothing'

There is one super-power that banks have which a casino does not. Every casino chip that circulates around the blackjack tables is 'backed' by US dollars held in possession by the casino. Banks, by contrast, are able to issue chips far in excess of the state money they hold. In other words, banks can issue a lot more promises for state money than they have in state money reserves (or they can issue these promises and fill in the reserves later). People sometimes call this 'fractional reserve banking', but the more accurate term for it is *credit creation of bank-money*. It is a mind-bending topic for many, and perhaps for that reason is widely misrepresented. But let us try to understand it, because it will help in our understanding of financial politics going forward, and it also provides inoculation from some of the more hysterical condemnations of banking.

The root of the misunderstanding lies in the fact that many textbooks still use the faulty 'cloakroom' image of a bank, in which deposits are not seen as *bank promises issued out*, but rather as *state money put in*. Notice that within this framework there is only one form of money (almost like a casino where cash circulates but where no chips are found). Within this paradigm banks are presented as intermediaries who store or move state money around, and who lend out state money they have taken in from depositors to borrowers. This 'one-form-of-money' idea leads to common misconceptions, such as those found in statements like 'The bank takes your money, but then lends it to someone else, which means your money is not actually there!' This is a bit like saying 'A hundred coats were deposited into the cloakroom, but the clerk has only kept ten of them. The rest have been lent out!'

I intensely dislike the cloakroom metaphor because it is extremely misleading, but let me briefly work with it in order to set the record straight. If a cloakroom were truly like a bank, I would put in my coat, but the cloakroom would take ownership of it and send it away while issuing me a *promise-chip* – recorded digitally – saying that I can come and get a similar one back any time. I now own a promise-chip, while it owns the coat. Imagine now, someone else approaches the counter with *no coat*, but nevertheless asks the cloakroom to issue them promise-chips for three coats. The clerk assesses them and says 'Yes, but only if you sign this agreement to bring me four actual coats in a month.' The agreement is signed.

The cloakroom gets a promise for four future coats, while the person is granted three current promise-chips, which are recorded digitally in the same system as my promise-chip is.

This is how bank 'lending' works. Notice that the cloakroom did not lend coats to the person, but rather simply *issued new promise-chips* to them, which is what they issued to me. There are two classes of things here: coats and promise-chips, and the main difference between me and the other person is that I obtained promise-chips by putting in a coat, while they obtained them by promising to *later* put in coats.

The result is that the two classes diverge: the number of promise-chips issued by the cloakroom jumps into excess of how many coats it owns. I the 'depositor' own one promise-chip, while the 'borrower' owns three, and the cloakroom owns one actual coat (taken from me), plus a promise for four coats (taken from the borrower). If you add those numbers up you will notice that the things they owe equal *four*, while the things they own – in theory and in future – equal *five*. Provided that the cloakroom can manage the process of chip redemptions (people trying to get *actual* coats) and provided that the person actually delivers four real coats in future, the cloakroom will end up with a whole extra real coat without owing anybody anything.

Thus, contrary to popular belief, banks do not 'lend state money' to people who ask for loans. To return to the more accurate casino metaphor, they simply issue out digital chips to people who ask for loans, and in return extract loan agreements from them. In more technical terms, banks expand the short-term IOUs they issue as a way to build up a war chest of long-term future loan agreements that are worth more than the IOUs they issue to get them. Provided they can manage the redemption processes and risks, this creates a 'pressure gradient' in which state money will be attracted towards the bank over time, enabling it to log a profit for its shareholders. As a depositor, your role is not to 'give the banks money to lend out'. Your role is to boost their reserves against which they issue chips.

This is 'credit creation of bank-money': the act of banks issuing chips in order to harvest loan agreements expands the money supply. In the UK, over 90 per cent of the money supply takes the form of these chips, and the entire digital money system works through their reassignment. All the politics of digital money, from surveillance to exclusion, stems from the fact that these bank chips are now becoming crucial to our lives, which means all manner of financial institutions are getting between us and our day-to-day transactions with each other. Let us take a quick tour of those different players by returning to my osteopath John.

'Moving' chips

Remember the question John asked: 'Cash or card?' We can now decode his request. He means 'Pay by physical state money or digital bank chips?' But, while the process of handing over a cash token is a simple physical act, the process of transferring bank chips is more complex. This is because they are assigned to accounts recorded in a fixed location – a data centre – and unless I can physically dig up the data centre and transport it, this money will never move. My bank, however, can *edit* my account in that data centre to re-assign the chips to John. Digital money 'movement' is just promise-editing.

If John banks at the same bank as me, it is as though he is in the same virtual casino. All the bank needs do is retract chips from me and re-assign them to John. If he banks at a different bank, though, it's as though he's in a different virtual casino, and this is where things get trickier.

Think of each bank as having its own *sub-currency* – in the UK, a Barclays customer is granted Barclays chips, whereas a Lloyds customer is granted Lloyds chips. Lloyds runs a private fiefdom of around 20 million UK accounts, so there is a huge amount of chip movement going on between Lloyds chip-holders *within* the fiefdom, but if one of those customers seeks to pay a Barclays customer, two fiefdoms are now involved. Because digital chips are attributed by a particular issuer to a particular account, they cannot be 'wrenched' out of that context and 'thrown' into another – a Lloyds chip attributed to a Lloyds customer cannot creep out of the Lloyds datacentre, run 200 kilometres to Barclays's datacentre, and creep into it to attribute itself to a Barclays customer. This is the principle of 'non-seepage'.

If I bank with Lloyds and John banks with Barclays, Lloyds must retract chips from me, while instructing Barclays to issue new chips to John. This, though, puts Barclays on the hook, so Lloyds must compensate Barclays by transferring digital state reserves to them at the central bank. This takes us back into the monetary centre of gravity – the central bank data centre. Just as we can reassign bank chips between each other, banks can reassign central bank reserves between each other (a process called 'settlement').

You might think of the central bank as the king who adjudicates between a series of commercial bank 'dukes'. The king does not want to be bothered with requests for adjudication until the dukes have resolved most differences among themselves. There are thousands of cross-bank customer chip transactions being proposed by those customers every minute, and it is too much hassle for the dukes to convene at the court of the central bank every time two of their lowly customers interact. Lloyds might have thousands of its customers requesting transfers to thousands of Barclays customers, and Barclays might have thousands of its customers seeking to pay thousands of Lloyds customers. Thus, rather than separately settling each one, they authorise the transactions, tally them up into bundles, and then superimpose those bundles over each other to cancel them out. If the 'Lloyds to Barclays' bundle is £100 million, and 'Barclays to Lloyds' bundle is £99 million, one duke need only reassign £1 million in state money at the central bank to the other.

The central bank likes this 'noise cancellation', as huge volumes of cross-payments can be made at the commercial bank level without it being disturbed. It stands aloof and quiet, only receiving sporadic messages saying things like, 'Re-assign £4 million from HSBC's reserve account to Barclays reserve account.' All transactions are psychologically anchored to this central bank level, but because of the cross-flow, not much actual state money ever has to be transferred.

International chip transfer

Just as bank sub-systems cannot seep into each other nationally, so national digital money systems cannot seep into each other internationally.* The definition of a digital British pound is 'a promise-chip recorded in the physical data centre of a British bank connected to the Bank of England'. It cannot be wrenched from that context and thrown over the ocean into another – digital British pounds cannot ooze down the road in Kuala Lumpur, intermingling with digital Malaysian ringgit. They are not floating units in space.

So how does digital money 'flow across borders'? A newspaper headline might proclaim that 'US companies pour billions into China', but these 'flows' are database accounting operations. While bank dukes can resolve their local differences via their local central bank king, there is no global 'emperor' to settle international scores, so national banks are forced to set up straddles to access a foreign system. They do this by opening special 'correspondent' accounts

* The countries of the Eurozone are an exception to this because their national systems fall under a common central bank

with each other across jurisdictions. Imagine virtual grappling hooks that tether their national systems together. Bank of America, for example, might open an account with the Industrial and Commercial Bank of China, and vice versa, providing a two-way link between the separate national systems. Over time, these two-way links have grown into a dense fabric of relations via which international transfers are resolved.

If I request a transfer to somebody at a Kenyan bank, our banks either must have direct accounts with each other, or may need to hop through a few degrees of separation by using another bank that has accounts with both. The transfer process sets in motion a cross-border domino chain – pound-chips leave my account and get credited instead to the UK account of the Kenyan bank, who – in a separate action – issues Kenyan shilling-chips to the person being paid thousands of kilometres away. The result is that my pound account goes down, while the Kenyan bank's pound account goes up, while the recipient's shilling account goes up.

The money does not 'jump' from the UK to Kenya. Rather, the international payments system involves foreign banks locally accumulating the chips of local banks. This process can be reversed when, for example, a Kenyan seeks to pay a UK resident: the Kenyan bank removes shilling-chips from them, gives up UK pound-chips it previously held, and has them transferred to the bank account of the British recipient.

To co-ordinate this, banks must communicate, but – given that there are thousands of banks internationally – it can get convoluted. This is where SWIFT – the Society for Worldwide Interbank Financial Telecommunication – comes into play as (in its own words) 'the global provider of secure financial messaging services'. The term 'messaging' sounds a bit like WhatsApp, but while SWIFT is a private communications hub, it is not for idle gossip. The only thing that moves in a digital money system is messages, and each SWIFT communication is a deadly serious legal order to initiate account editing. Its system of Bank Identification Codes (BIC) provides a way for banks to do that in a standardised format across borders, via SWIFT's datacentres.

These orders could be sent via a different messaging network, but SWIFT already has some 11,000 banks using its system globally, and to convince those banks to switch would be difficult (much harder than convincing all your friends to simultaneously switch to a different messaging platform). And, as Iran has learned, getting excluded from this network is a sure-fire way to strangle cross-border payments, meaning SWIFT has become enlisted as a tool of geopolitical influence.

Communicating with the banking cartel

The objective of any digital payment is to get banks hosting two accounts to edit those to give the illusion of money hopping out of one into the other. To initiate this we must ask them to do so. I cannot, however, stand outside my bank's data centre and shout this request. I must translate the request into digital code and send it to them from afar. Banks provide us with a limited number of 'transmission posts' to do this either directly or indirectly. The classic direct channel is the branch, where I speak to a bank employee who can input my requests via their computer, but I can also phone such a clerk or send them postal requests. And these communication methods have been overtaken by Internet banking pages that allow me to send digital messages directly to bank datacentres.

While Internet banking is fine for predictable large payments like rent, when we are out in the street it's a different story. Imagine arriving at the counter of a bakery, taking out your laptop, asking for the Wi-Fi code and the store's bank details, initiating a transfer and then waiting for it to log into its account to see confirmation as a queue of irate customers extends behind you. This is why *indirect* channels exist: cheques, for example, are written orders handed out to a seller, allowing them to collect the payment from your bank later. The major indirect channel we use now, however, is card payments.

The card networks, like Visa and Mastercard, were ground-breaking because they created a system for indirectly messaging your bank in situations when manual transfers are difficult or insecure, such as when you are in a store (or, in the online world, an e-commerce store). Rather than asking wandering buyers to use their own transmission posts, the system gets *merchants* – who are more fixed – to establish them.

Card schemes are like co-operatives that make profit for their co-ordinators (like Visa) and their members, which are banks. Member banks issue cards to wandering buyers like myself, while setting up merchants like John with transmission posts to allow me to initiate payment requests via the network maintained, in my case, by Visa. These transmission posts include physical point-of-sale (POS) terminals that you insert your card into, but also digital 'payments gateways' installed on websites, which you get sent to when making an e-commerce purchase (think of these as the digital equivalent of a POS terminal).

A modern bank card is like a little computer, a wafer-thin laptop that I can store in my wallet. Its job is to shout a digital message through the POS terminal, sending my details via John's payment processor into VisaNet, Visa's behemoth data centre system. My request will be one of around 5,000 arriving in the same second, initiated by customers of some 15,000 banks and payments processors that plug into this system. Visa's data centres can route up to 30,000 simultaneous transaction requests per second, but it will not reveal where its primary datacentre is, saying only that it is 'along the Eastern seaboard'. The facility is of such strategic importance that it has emergency power generation facilities to last nine days, and major security requirements too. According to *Network Computing* magazine:

> The roads entering the complex have hydraulic bollards that can shoot up fast enough to stop a vehicle traveling up to 50 miles per hour dead in its tracks. (The road is too curvy to drive safely at higher speeds.) Visitors must pass through a security gate, be cleared by roving security teams, and then be subjected to a biometric scan before being admitted.

This data fortress routes the message to my bank, the digital equivalent of saying, 'Brett wishes to pay an osteopath – does he have chips available?' My bank checks, and then relays a payment authorisation back along the same chain. John's terminal beeps and displays a message: 'Payment authorised'. I walk out, and two banks initiate a settlement process.

The basic principles of digital payment are straightforward: there are banks, messaging platforms between them, and messaging devices for us. From here you can understand most payment 'innovations', which normally entail building a layer on top of the same system or creating a new way to message the banks directly to bypass the card networks, or to augment those card networks. ApplePay and GooglePay, for example, are just new ways to send messages into the same old system, turning my phone into the equivalent of the card (with the side effect that Google now gets a new data stream of my payments activity).

The problem of surface appearances

The question 'What is money?' – like the questions 'What is art?' or 'What is the meaning of life?' – can be debated at length with no clear conclusion. I tend to bypass this discussion by describing our current monetary system: money is a composite system of state and bank IOUs, activated within legal systems set within political systems, which in turn come to act as economic-network access tokens among vast interdependent networks of people. Once kick-started into life, money solidifies its strength through network effects, the situation of millions of us being interdependent, in the context of institutions that reinforce and shape that interdependence.

While states and banks are the core institutions, many 'non-bank' systems – such as PayPal, Venmo, WeChat, M-Pesa and Paytm – can plug themselves into the banks as add-ons. If banks take your state money and issue you digital bank chips in return, players like PayPal take your bank chips and issue PayPal chips to you in return. Numbers in your PayPal account are therefore third-tier corporate IOUs for second-tier bank IOUs for first-tier state IOUs.

These systems are chained together with accounts: I might have an account at PayPal, which has an account at a commercial bank, which has an account at a central bank, which has accounts with other central banks (while using the specialist services of mega-institutions like the IMF and the Bank for International Settlements). Once you reach this international realm, the politics get very big. Central banks, for example, give each other massive reciprocal credit lines to secure emergency access to each other's currency – for example, in 2013 the European Central Bank set up a 'swap line' with the People's Bank of China, giving it access to 350 billion yuan while giving the Chinese central bank access to €45 billion.

Given the sheer power of the banking sector, I like to visualise these systems as governing our interactions *from above*. In the previous chapter I developed a metaphor of the giant in the mountain – which does convey the idea of a power that looms above – but modern monetary institutions have transcended this, and are now closer to cloud-cities in the sky hovering above each country. They are tied together internationally via messaging systems like SWIFT, card networks run by Visa, and payments clubs like SEPA. Crucially, within that international system there is a *hierarchy of cloud-cities*: smaller national systems often route relationships between each other through the US banking system, with the Federal Reserve on top, which is why the US dollar is referred to as a global reserve currency.

We send millions of payments requests up to this structure, but our bustling interactions are far from our national apex, and even further from the international apex. From the ground level, those institutions seem distant and abstract, but central banks' attempts to expand or contract the money supply are an attempt to exert indirect influence over us, via the commercial banks (this certainly is indirect, because central banks have limited control over the overall money supply, given that it is significantly made up of commercial bank chips). Monetary policy seeks to alter the volume of impulses that fly through the monetary nervous system, whether to mobilise us to produce, or to slow us down. This leads to all the battles waged between factions of monetary economists (like Keynesians or Monetarists), who fight over the relationship between the quantity of money units issued and the amount of real things of value produced by people in the economy.

Monetary policy talk can be bewildering, but being a money user in general is bewildering. Because the economic networks we are enmeshed within are vast and spread out, it's difficult to zoom out far enough to see the structure of the monetary systems that hold them together. For most of us, the readily visible aspects of the system are physical artefacts. So when a journalist writes an article about money the accompanying image might be from a stock photography site like Getty Images, which illustrate 'money' with pictures of dollar bills or pound coins. These images do not show us money in a general sense. They show us physical state money tokens – just one part of a much larger system.

There is no imagery of bank data centres or digital chips, and these don't come up when we type in 'digital money' either. Typing that phrase into a Google image search gives us images of physical cash tokens dissolving into data, or cash flying through wires, or cash emerging out of a computer.

These popular images imply that cash should be thought of as a generic symbol for 'money in general'. But the fact that we struggle to define a boundary between cash and digital bank chips has serious implications. For example, in 2016 the Panama Papers were leaked: 11.5 million documents chronicling the shady dealings of the off-shore finance industry. The official leak website – covered by countless media organisations – came with a strapline that said, 'Politicians, Criminals and the Rogue Industry That Hides Their Cash', accompanied by pictures of dollar bills floating in stormy winds. The offshore companies uncovered, though, operate through secretive bank accounts that use digital bank-money, not state cash. For those who did not look past this headline, the Panama Papers' casual labelling of bank transfers as 'cash' unwittingly hides the banking sector and this parallel system that it runs. In this way people begin to associate cash with crime, while overlooking the fact that a great deal of large-scale financial crime occurs using straightforward bank transactions.

Much of our public imagery seems intent upon ignoring the reality of bank-issued digital chips, even as they take over. In Chapter 1, I characterised the financial sector as the nerve centre of global capitalism, noting that this sector increasingly looks forward to a cashless society, such that it can extend its oversight across the entire 'nervous system'. It suits the banking sector that digital money be represented as a 'cash-like' digital unit. It suggests their chips are merely an upgraded version of cash. Cloaked under the reassuring imagery of physical tokens, their systems avoid our scrutiny.

Cash tokens were once a key force by which the state spread its logic, but they have outlived their usefulness to the corporate class that has emerged under those states. Cash tokens might be issued from on high, but once they are among us, they are – literally speaking – down to earth. They maintain localisation, moving between people in physical contact with one another. Bank chips, by contrast, are invisible, distant, detached and under the control of institutions that are far easier for mega-corporations like Amazon to work with.

I call these digital chip systems 'cloudmoney' because they all rely upon the cloud – data centres that we interact with from afar. Cloud imagery is useful but dangerous, because we run the risk of imagining clouds as floating entities free from material reality, when in reality the cloud is a hidden complex of compounds surrounded by electrified fences and armed guards. I can take a match to a banknote and send it up in flames, sacrificing a state promise, but I cannot burn digital money by burning my credit card, phone or computer. The only way to burn bank chips is to find the data centre, break in and engage in industrial-scale arson. Compared to the new private cloudmoney units that live within these gated data centres, cash tokens now look positively friendly.

5

The Bank-Chip Society

Picture yourself winning big at a casino blackjack table, raking in a pile of chips, then heading to the cashier to convert those to cash and leave. The cashier looks up and says, 'Sorry, we don't redeem those any more.' They let you into their casino, but now are trying to stop you exiting it. Would you be angry?

Something analogous is happening with banks. In countries like the UK they are quietly shutting down ATMs and branches, which are ramps into and out of their systems. According to data from the British Bankers Association and the Office of National Statistics, between 2012 and 2020 the number of UK bank branches declined by 28 per cent, while ATM numbers declined by 24 per cent between 2015 and 2020. Perversely, banks are using their success in absorbing people into their systems as a justification for closing down future opportunities for people to exit. We've seen that we live under a dual monetary system – with state cash and digital bank chips – but in the 'cashless society' one of those options disappears, leaving bank chips as the only choice. 'Cashless society' is a euphemism – as uninformative as calling whisky 'beerless alcohol' – but the financial industry likes the phrase because it draws attention to something that is absent, rather than to something that is rising to power. Imagine how much harder it would be to market the 'bank chip society'.

Remember that banks retire our cash from circulation, 'dematerialise' it into reserves they own at the central bank, and then issue us chips in their data centres, but they are also supposed to run the reverse process. When we request state money, they have to destroy chips, and re-materialise reserves into cash that they send out via ATMs and branches. But banks have begun to present ATMs as a helpful but outdated public service that they are encumbered with running. It is like a casino presenting the redemption of their chips as a charitable service they offer, rather than a legal requirement. Banks now present themselves like this while slowly closing down the cash infrastructure.

This makes the chief competitor to their chips – cash – harder to access, which in turn makes cash seem less convenient, which in turn pushes more people towards their systems, an influx that is then used to justify closing down further cash infrastructure.

That infrastructure includes bank branches, which are not only being progressively closed, but are also having their cash-handling capacities reduced. In Sweden, for example, many branches refuse to take in cash. So if you are a shop owner who accepts cash, you have to travel further to deposit it at the end of the day. This pushes merchants towards refusing cash – or 'going cashless' – which in turn shuts down opportunities for customers to spend cash. We end up with cash being hard to access and hard to spend – a recipe for an implosion in cash usage. In other words, banks are in a position to engineer a self-fulfilling prophesy, and the strategy is succeeding.

Worldwide Google Trends data shows a significant increase in searches for the term 'cashless' since 2014. The term, however, now refers both to the general case of a society without cash ('cashless society'), and to the specific case of a single digital payment (a 'cashless payment'). In logic, to give identical names to two different things can lead to an *equivocation fallacy*, in which the separate meanings pollute each other. This is happening when, for example, the French government calls its bank digital payments plan the 'cashless payments plan'. It is entirely possible for cash and digital bank chips to *co-exist* but, given that the latter go under the name of 'cashless payment', it is impossible to speak about them without implying the end of the former.

With bank-money disguised under the term 'cashless' it is no wonder that the issue is so hard for people to discuss clearly. With older generations, making a distinction between 'money outside the bank' (cash) and 'money in the bank' (bank-chips) still holds, but this distinction risks becoming lost on those coming of age in a situation of digital dominance. For these younger generations, bank-money has gained psychological ascendancy, and they are increasingly swayed by the argument that bank chips could detach from the state money system entirely, requiring no access to cash at all. In this 'great forgetting' the ATM becomes seen as an antiquated relic, rather than a means to redeem the promises for state money issued by the banking industry. They are imagining a world in which there is no such thing as 'money outside the bank'.

This will have serious political, economic and psychological consequences. For example, when a bank looks as though it could fail, people – historically – panic and rush to redeem its chips for state money in what is known as a 'bank run': they queue up at ATMs and branches trying to 'get their money out'. But what happens if there are no ATMs or branches? Well, you are going to have to log onto your account and try to flee your bank by making transfers to another bank, perhaps by asking your friend if you can temporarily transfer to their account. But what if the whole banking sector is in crisis?

Get in sync with the future

This possible future inability to escape the confines of 'money in the bank' (or, more accurately, money issued by the bank) should be seen for what it is – an *enclosure*. Yet the mainstream narrative characterises the bank-chip society as an inevitability driven by ordinary people who wish to be enclosed into benevolent digital systems. Finance futurists assert that this march towards digital enclosure is 'natural progress', an unstoppable rocket fuelled by Generation Z and millennials (people born after 1981) who 'demand digitisation'. Digital payments companies rely heavily on this story, implying that if you don't get on board, you'll be left in its dust.

PayPal, for example, drew upon this narrative in 2016 when it covered the London Underground with its 'New Money' campaign, replete with images of young middle-class up-and-comers. In the campaign PayPal addresses the viewer as the excited recipient of a wished-for technology. Its 'New Money is Here' slogan sounds as though it was uttered by a delivery man bringing you something you had requested.

The French theorist Louis Althusser referred to this technique as *interpellation*: a method of programming ideas into people by addressing them as if they already agree with the ideas. 'Cyber Monday', for example, is no more than a creation of the National Retail Federation, which reverse-engineered its existence by issuing a press release saying, 'Cyber Monday Quickly Becoming One of the Biggest Online Shopping Days of the Year.' All the resulting commercial slogans – 'Cyber Monday is Here'; 'Are You Ready for Cyber Monday?' – hail you as someone who accepts that the day is real, and conjures imagery in your head of millions of others who believe it is real, eagerly waiting for it. All digital payments companies – like PayPal – use similar techniques, designed to project a sense that 'new money' is a real phenomenon, being carried along by a movement of empowered people.

Trying to maintain a critical outlook in a society dominated by this corporate messaging is challenging because it saturates our environment, leaving one feeling like an agnostic amid a well-funded religion. There are no billboards saying, 'Cash is a public utility that should be protected,' because cash is not attached to a private sector company that will promote it. Rather, it remains undefended as London's professional community wait for trains on the platforms of the London Underground, staring at billboards on the opposite wall where payments companies address them as post-cash converts. You may be a person who feels rebellious at the inauthenticity of this messaging, but even then its saturation can lead you to imagine that everyone around you sees it as common sense. This in turn conveys a more hostile message, namely: *change will happen regardless of whether you're involved or not*. If you reject the 'new money' you are clinging to the past, out of sync with everyone else.

In Chapter 2 I laid out some initial reasons for why we should be suspicious of a vision that presents the bank-chip society as driven from the 'bottom up' by people who desire it. It does of course take two to tango – banks and quasi-banks like PayPal cannot overtly force people to use their systems – but how have we been convinced to open ourselves to this takeover?

Attacking the bicycle of payments

In 1898, the Winton Motor Carriage Company published the world's first automobile advert. 'Dispense with a horse,' it said, 'and save the expense, care and anxiety of keeping it.'

An early term for these motor carriages – or motor cars – was 'horseless carriage', suggesting that the innovation was unencumbered by a

DISPENSE WITH A HORSE

and save the expense, care and anxiety of keeping it. To run a motor carriage costs about ½ cent a mile.

THE WINTON MOTOR CARRIAGE

is the best vehicle of its kind that is made. It is handsomely, strongly and yet lightly constructed and elegantly finished. Easily managed. Speed from 3 to 20 miles an hour. The hydrocarbon motor is simple and powerful. No odor, no vibration. Suspension Wire Wheels. Pneumatic Tires. Ball Bearings. ☞ *Send for Catalogue.*

Price $1,000.   No Agents.

THE WINTON MOTOR CARRIAGE CO., Cleveland, Ohio.

previous limitation. Nevertheless, they encountered resistance, especially in Europe, because people were attached to their horses or could not afford motor cars. Car enthusiasts saw this as stalling an inevitable positive change. Picture them trying to overtake a horse cart on a country road while shouting, 'Make way for the future!'

Similarly, digital money promoters present cash as the horse-drawn cart of payments, claiming that it only survives through the stubborn nostalgia of laggards. They see digital money as an 'update' to cash and, much like the term 'horseless carriage', the term 'cashless payment' as implying that some previous hindrance has been shaken off. And why hold onto an inferior system?

The 'horseless' motor carriage was indeed an upgrade to the horse-carriage: both had a power source, wheels and a carriage, and both used the same roads. Digital bank chip transfers, however, are not an upgrade to the state cash system, given that the latter under-pins the former (in much the same way as casino chips are not an upgrade to the cash they promise you access to). In fact, the only thing that digital bank transfers should be seen as upgrading are non-digital bank transfers. A 1920s cheque was a physical order sent to initiate the editing of a bank account, whereas a 2022 smart-phone payment is a digital order sent to initiate the same thing. It is the cheque that is the 'horse cart'.

Partisans with an anti-cash agenda paint cash as an impediment – blocking the road for the fast cars seeking to overtake it. Yet there is no conflict in maintaining both systems. The closest transport analogy for cash is the bicycle, and 'going cashless' is like closing down bike lanes that run parallel to roads in a city of cars.

When a product or service is upgraded, people no longer want the old version, but when a product or service runs in parallel, they are very likely to wish to retain both. Email, for example, destroyed the fax machine but not the postal service because, despite the over-lap in functions, post still does things that email cannot. We may be indifferent to the death of fax machines, but an outcry would accompany an attempt to close down postal delivery. Similarly, lifts have their use, but no responsible property developer would ever *only* install a lift without having emergency stairs. Keeping parallel options supports the principle of resilience through diversity, and it's the reason municipalities promote multi-modal transport sys-tems. The bicycle *preceded* the car, yet is more popular than ever, given that it eases problems associated with cars, such as air pollu-tion, congestion and the stress and poor health of urban life.

Similarly, a diverse and resilient payment system requires non-digital and non-bank systems. Our payments system, however, is being driven towards the mono-modal: a plethora of digital pay-ments companies might give the illusion of diversity, but they are all built upon the same underlying banking oligopolies. Because this industry is seeking to win power for one mode, they behave like old automotive lobbyists scaremongering about the dangers of alternatives while ignoring the scores of motorcar accidents.

Indeed, the automotive industry – starting with the Winton Motor Carriage Company – only advertised the narrow, individual and short-term benefits of cars ('Save time and expense') and left it to governments to deal with the longer-term collective social fall-out via road rules, congestion charges and fuel standards (and bike lanes). A similar dynamic exists with digital payment: just as car adverts are set in open countryside roads, rather than in polluted traffic jams or cordoned off crash sites, so adverts for digital payments don't say, 'Enjoy the speed, convenience, surveillance, cyber-hacking and critical infrastructure weaknesses that our platform brings.'

Despite the increasing hold of the story of cash as a 'horse cart' to be dispensed with, it remains uncomfortable to many, who at some intuitive level get a 'bicycle-like' feeling from cash: it may not go as fast, but it's great for short outings, requires less maintenance, is more inclusive and certainly comes in handy when the other system gets jammed up. A digital bank account depends on getting access to 'the system', and on that system being maintained – but what type of person is prepared to place their full trust in such systems? Let's turn to that question.

Horizontal informality vs. vertical formality

When I was eighteen, I busked on the New York subway with my guitar and a harmonica strapped around my neck. I wasn't exactly Bob Dylan, but a few passers-by flipped me some coins with a smile, enough for me to buy a slice of pizza. Busking involves gifting music to strangers, who occasionally hand over a token of appreciation in return. Historically, this token took the form of cash, which can be thrown spontaneously. Busking is a great example of a small-scale situation in which parties casually relate to each other without need or desire for a formal mediator – but that informal culture is under threat. Now street rappers are being told that they must first have a contactless payments device, and that Mastercard will be taking fees from them as they busk.

The informal ethos that is under threat is not limited to busking. The same can be said for pub quizzes and home poker games, where people have traditionally pooled their entry-fee money into a beer mug, or a typical Polish wedding, where cash is pinned to the bride's dress in exchange for a dance. In devotional Sufi qawwali concerts, audience members approach the stage and theatrically throw cash to the performers, while small-scale farmers in rural Germany (and many other countries) set up unattended roadside stalls with wooden 'honesty boxes' to gather coins from those who take the fresh produce stocked there. Countless millions across the world live with a similarly informal economic outlook, and the use of cash – which enables them to bypass formal institutions – is central to that world view.

This was reiterated to me when I visited New York in 2017 and found myself standing on a road between two establishments. The first was a family-run Asian restaurant. On the table a sign read: 'If you are able, paying with cash is better for mom and pop shops like us. Credit card companies charge a fee with every transaction.' Directly opposite was a chain restaurant-café that targeted health-conscious young professionals with organic salads and kombuchas. On their door was a different sign: 'No Cashew Money: This Sweet-green no longer accepts cash – download our iOS app to get your fix. Head to bit.ly/sgcashless to learn why we went cashless.' I visited the link, and was greeted with a blog post titled 'Welcome to the Future—It's Cashless'.

These venues were separated not only by their payment

preference, but by a cultural barrier. The first sign, homemade, put forward a straightforward request. The second had been designed by a graphic artist, featured a corny attempt at a wacky pun, and carried the assumption that it was weird not to be using an iPhone. In the first establishment I could ask the family why they preferred cash. In the second I would be directed to a website put up by the owners of the chain – the employees would just shrug their shoulders. The first was a down-to-earth diner where anyone would feel comfortable, whereas the second establishment was run for those who want to be on the up – professionals with laptops who gather to have meetings about start-up strategies, events organisation or a new fashion label. Data from several studies shows that cash usage is lowest among those with higher incomes and education, but you don't need to be a social scientist to see the obvious class divide in payments choices. Venture to any trendy café where the clientele have good credit ratings, and you will quickly see that digital payment thrives among social climbers who see themselves as sophisticated.

This is not often reflected in commentary about the future of money, because commentary tends to be a speciality of people in higher-status social circles. Take, for example, Kenneth Rogoff, the high-profile author of *The Curse of Cash,* formerly head economist at the IMF, and now a professor at Harvard. His anti-cash opinions get easy access to politicians, high-profile academics and the media. Others, like Richard Thaler – who won the Nobel Prize in economics for his work on nudge theory – publicly praised India's Modi government for clamping down on cash. These are the people conditioning our views on cash, but they do not represent its culture.

A social worker once suggested to me that a person's social status can be ascertained by the level of trust they place in the police treating them fairly. The marginalised young people he worked with were often automatically assumed criminal by authorities, which meant they had no reason to trust that those authorities would treat them fairly. This intuition is widespread among people in lower socio-economic classes, who often have uneasy relationships with official institutions – which include banks – perceived to be run by and for social elites. High-profile economists, like the average Sweetgreen customer, seldom experience this distrust of institutions. Are these the people most likely to view 'cashlessness' – which relies upon constant interaction with formal financial institutions – as unproblematic 'progress'?

The Better Than Cash Alliance does not think so. From its New York office, its directors characterise cash as an outdated and dangerous drag on the economy. They assert that it is those living in informal economies around the world who yearn to leave it behind, and upgrade to the lifestyle of the digitally connected urban professional.

The ambiguities of being 'banked'

Organisations like the Better Than Cash Alliance position themselves as players within the broader field of *financial inclusion*. Practitioners in this field often use the term 'unbanked' (or 'underbanked') to refer to those who do not use formal financial institutions in their day-to-day lives. Someone living in a small rural town might think it normal to ride a bicycle to an outpost store where they use cash. Through institutional eyes, however, the transport could be characterised as 'Un-Ubered', the commercial setting as 'Un-Amazoned', and the person as 'unbanked'. It's a descriptor that carries a value judgement: it implies that it is preferable to be absorbed into the banking sector.

Many financial institutions historically avoided engaging with poorer people because they were seen as relatively unprofitable. This is not always true – banks can make a killing off vulnerable people, as they did with sub-prime mortgages in the early 2000s – but offering current accounts to low-income individuals is relatively unattractive. This is because financial institutions are attracted to *risk-adjusted profit*: the ultimate 'irrational' business line is one in which they take high risk and expect low (or no) monetary return, and the ultimate 'rational' one is that which yields high returns with low risk. When deciding upon which customers to offer accounts to, they carry out risk-return calculations, considering how costly it is to manage those accounts, and contrasting that with how much revenue they can expect to make from the customer by getting them to pay fees or take loans. Finally, they consider the risk the customer exposes them to.

Banks have traditionally bent over backwards for rich aristocrats and merchants, in the knowledge that providing them with accounts was likely to yield good returns over time at relatively low risk. People on a lower income often do not fare as well in this same calculation, because the bank's fixed costs for providing an account remain, but the customer's smaller transactions for lower amounts do not yield as much in fees or interest.

The economic equation, however, becomes more attractive if fees and interest can be scraped off large volumes of poorer people without those people coming into a branch or demanding too much service. This is why mobile phones were hailed as the Holy Grail for banking in the developing world. Financial inclusion professionals who seek to incorporate poorer people into the banking sector recognised that phones could lower the fixed costs of providing services, which – when passed through a risk-return model – made serving poorer people more profitable (while also allowing banks to integrate financial data with other data collected on phones to build a profile of the customer).

Almost all financial inclusion initiatives present digital technology as a great leap forward that will enable 'the unbanked' to get banked. Not mentioned, however, is that the economic risk-return equation is only half of a bigger equation. While banks may not like poorer people unless they can make dealing with them profitable, poorer people have often had no natural reason to like banks either. One part of this is practical: historically, the average size of their transactions was so small as to make writing cheques or requesting bank transfers an unnecessary or even embarrassing process (especially in situations where they might only buy essential goods within a small radius from where they live).

Another part is political. I was a boy at the tail-end of the apartheid regime in South Africa, notorious for its policies of racial discrimination. At this time my parents opened a special children's account for me at Standard Bank, one of the country's most prominent financial institutions. I remember the branches being full of white people, while black people stood outside. Gradually, as South Africa moved into its post-apartheid phase, the number of black customers slowly increased, but those who were illiterate were prone to being treated with condescension. For an elderly Zulu man who has spent his formative years as a labourer for an apartheid mining corporation, there is no reason to feel trusting towards financial institutions.

This same pattern is found the world over: not only are banks associated with elite classes, but they are intimidating and have frequently abused poorer people. When I was based in the UK between 2008 and 2021, the country was repeatedly rocked by banking scandals in which low-income people had signed contracts written by biased lawyers to mis-sell them products they did not fully understand. And the history of finance is scattered with tragic scenarios in which bubble-like financial products are sold to the poorest of society at the very moment those products are about to collapse.

That financial institutions are not widely seen as a friend to the poor is reflected in the payments landscape. Cash economies inevitably form on the peripheries of society, such as previously tribal regions with rural smallholders or indigenous people. Cash is used by informal fish marketers in Maputo, back-street hairdressers in Mumbai, and self-employed Andean craft merchants. In richer economies there are pockets of society that have always relied on cash, which is associated with hour-by-hour employment as opposed to the salary-based payrolls of the professional classes. It finds a home with marginalised ethnic minorities, rural smallholder farmers, small-town misfits, drop-outs, free spirits, rugged frontiers-people and struggling artists. The ethos extends to all the low-status employees who run the back end of cities: the street sweepers, the janitors and the security guards standing outside J. P. Morgan's offices (along with the sex workers and drug dealers that the financial professionals within those offices occasionally like to visit).

Wanting to include vs. wanting to be included

Initiatives like the Better Than Cash Alliance promote the view that cash is a second-rate system for those who cannot get access to a first-rate one – and present themselves as champions who seek to *include* poorer people into the *better* corporate services they should be getting. But the power dynamics of this are complex. Let me illustrate through another personal example.

I was awarded a scholarship to study at the prestigious Cambridge University. Upon my arrival, however, a posh elderly professor confided in me that the scholarship was actually intended for black South Africans, but I had been given it because no black South Africans had applied. He offered various theories for why this might be the case, but as we drank port in a hall that looked as though it had come out of *Harry Potter* I developed a theory of my own: Cambridge is a symbol of the British Establishment and, by extension, British colonialism. The scholarship was supposed to promote inclusion into this insider club, but many black South Africans are likely to associate the club with exploitation. Perhaps this contributed to the lack of applications.

I had no direct evidence to support that hunch, but it is the possibility that counts, because it is a possibility that is seldom taken seriously in top–down inclusion initiatives more generally. Such initiatives, from the Indian Aadhar system to the myriad financial inclusion programmes pushed by international agencies (and their corporate partners), typically proceed from the assumption that those they seek to include *wish* to be included. This same assumption is found in development language, when 'less developed' people and nations are presumed to wish to join the club of 'the developed'. This 'club' is centred on major international cities like New York and London, where the wealthiest business leaders, most powerful politicians, coolest influencers and most buzzing in-group 'scenes' get access to the most advanced technologies, while controlling the narrative around technology and economic advancement.

It may be the case that some on the periphery do wish to be included into this insider club, but there will be many more who are indifferent or rebellious towards it. This is where we get back to cash. There are countless examples of people who state a preference for cash over the supposed upgrade to digital payment. I have collected the stories of many such people, from tattoo artists and fish sellers in London to El Salvadorian merchants, who assert that they want to physically see and feel the money they make. Regardless, in mainstream inclusion initiatives such preferences are assumed to be temporary anomalies: it is not that people actually prefer cash. It is just that they haven't yet had an opportunity to realise that they prefer digital payment.

But modern digital payment is under the control and oversight of globalised finance corporations, and to become dependent on that system is to enter their sphere of influence. The dark skeleton looming in the 'digital financial inclusion' closet is this: while there is a recognition that the global economy is deeply unequal both within and between nations, the urban professional classes (of San Francisco, Mumbai, Paris or wherever) largely do not question whether the digital economy they seek to expand breaks down existing hierarchies. Indeed, the goal of 'inclusion' could be to bring more people into the digital financial net but in a subordinated position.

Many people on the peripheries of the global economy intuitively understand that cash shields them from those in the core, with physical state money enabling participation in capitalist society while simultaneously offering the user some protection from its elites. For all its 'inefficiencies', cash has no hidden agenda, and is inclusive in a no-strings-attached way. The notes roam without concern for what neighbourhood they find themselves in. They can be rolled up to snort cocaine in a fashionista soirée, or be used to buy nappies in a corner store. The note is non-judgemental, a kind of 'everyman' serving rich and poor alike. For anyone who distrusts institutions that do not automatically protect them, cash offers some breathing room.

Let's face it. Cash gives space for back-street, small-scale, family-run-business styles of capitalism. Corporations might battle each other for dominance, but they are allied in their desire to conquer that style. They want to swallow up mom-and-pop shops and consolidate them into a major branded chain, or to displace an informal street market with a supermarket listed on the stock exchange. Cash, in other words, appears resistant to both the ethos and future development of corporate capitalism.

In Chapter 1 I noted that corporations use digital bank transfers between themselves – to facilitate large-scale bulk production and wholesale trade. They have, however, historically relied upon cash in the hands of people from low socioeconomic backgrounds as the 'last mile' form of payment to get, for example, an individual bar of Unilever soap to a factory worker in a Bangladeshi corner store. To the financial and corporate establishment, cash is a necessary evil, because it completes the profit cycle – which they like – but does so while blocking their institutions' attempts to automate systems, spy on customers, and discipline people through threats of exclusion. Cash also blocks an enormous amount of data from falling into corporate hands, and data – as all of us should know by now – is a hot commodity.

6

Big Brother. Big Bouncer. Big Butler

I first met Tyler Hansen at Symbiosis Gathering, a hippie festival in Oakdale, California, where I was co-running a space for hackers, artists and activists called the Hacktivist Village. Tyler was a classic 'burner' – a veteran of the legendary Burning Man festival – and had come to my rescue in Oakdale when our amateurish hand-built stage was on the verge of collapsing in the hot Californian gales. I thought I would never see him again, but two weeks later I was strolling down Sunset Boulevard in Los Angeles and walked straight into him.

Over a drink, he told me about the problems confronting the marijuana industry. In recent decades, US cannabis activists have made progress, and recreational smoking is now legal in nine US states, including California. However, while legalisation campaigners have won at the individual state level, US banks are regulated at a federal level, which means that, fearful of falling foul of federal laws, they block legal marijuana companies from opening accounts. As a result, the weed economy largely remains a cash economy.

Tyler wanted to explore possible responses to this, and invited me to participate in a 'Cryptocannabis Salon' that he hosted in a circus tent in the docklands of Alameda. It was a meeting between cannabis entrepreneurs and cryptocurrency pioneers, two groups walking the same fine line between innovation and illegality.

The legal cannabis industry is full of people once considered criminals who are now recognised as legitimate businesspeople. This status change takes adjustment. They still bear symbols of an underground lifestyle – like bling-studded leather waistcoats – while trying to learn the styles and mannerisms of professional business. A slick former banker – now the manager of a marijuana venture capital fund – came to the salon to do a PowerPoint presentation on the profitable opportunities of the industry; the Wild West of weed is a new target for Wall Street.

Tyler had requested that I follow with a talk called 'The War on Cash', but as I started speaking I could tell I was facing a tough crowd. The inability to obtain access to banking services jars with the outlook of the new professional marijuana entrepreneur. Many wish to gain access to the bank payment systems they have histori-cally been excluded from – as legal businesses, they must pay tax, but they find themselves having no choice but to pay it in bundles of cash. Through their frustration, some have become dismissive of cash. It was like trying to give a talk on the joys of cycling (and why the auto industry is dodgy) to a teenager who is tired of riding a bicycle and wants a car.

I stood firm, though, and reminded the participants of one key point: *without cash, you would not have survived long enough to get your industry legalised.*

Progressive social change is often marked by the legalisation of once criminal behaviours, examples of which include homosexu-ality, interracial relationships and – following Prohibition in the US – alcohol. Authorities attempted to cast drinking alcohol as a black-and-white case of 'doing wrong', and yet tens of millions of people continued to drink, suggesting that while they recognised its illegality, they did not see drinking as immoral. Rather, it was a 'grey area', and attempts to clamp down on grey areas have always resulted in driving them underground, where cash provides life support. Cannabis too relied on the cash economy while its advocates fought for its legalisation. Now that is paying off as the positive role it can play in many medical conditions is being recognised.

But if cash were to be fully phased out, this type of life support could be severely curtailed. Zealous attempts to choke black markets have the effect of simultaneously choking routes for creative deviance in a society. In a world where a citizen's every activity can be monitored via digital payments, the ability for law enforcement to locate and stamp out grey areas greatly increases. If you believe grey areas are a valuable interstitial zone where positive change can flourish, their removal should be seen as an attempt to freeze society into rigid zones of black and white. Alternatively, if you believe people will always establish new grey areas, in the absence of cash they will necessarily be sustained by the proliferation of new non-cash, underground means of payment, such as cryptocurrencies.

It may seem reasonable for law enforcement agencies to use whatever technological means are at their disposal to do their jobs, but as these means get ever more powerful, so their power to perform their jobs beyond prior expectations increases. Such overperformance can quickly extend into overreach: eighty years ago, no government ever *expected* to be able to finely monitor all payments, and therefore did not see it as a right. But increasingly, both states and payments industry lobbyists openly suggest that those who use cash should be treated with suspicion for their failure to convert to digital.

Having the technical capabilities to perform widespread monitoring is rapidly morphing into an expectation that these capabilities *should* be exercised. Because these shifts in expectations happen incrementally, we are susceptible to a process of 'slow-boiling a frog' – often we cannot detect our rights being eroded until it is too late. This slippage is a dangerous process, and privacy activists fight it, arguing that the new capacities to surveil people should not automatically be exercised. Let's take a closer look at the implications of bank digital payments on our privacy.

Watching from afar

Picture a sixteenth-century fishmonger. He would most likely have been well known by friends and family, and through reputation by a local network of associates, but poorly known by strangers further afield. Watching through a telescope from afar, an agent aboard a passing ship might have spied the man upon the docks and made an assessment. *Male. Sells fish. Strong. Seems to live in that cottage.* This is very rough and incomplete information.

Over time, companies and governments have developed more sophisticated telescopes and observation decks from which to obtain information about people, but even as recently as the 1980s they were still only able to capture data points that rendered a rough image. Various authorities would know things like: *bought a house in X neighbourhood, went to university at Y, employed at Z, single female aged 34 of roughly this socio-economic status.* In pixelated form, you would look a bit like a video game character from the 1980s – a rough two-dimensional caricature.

However, in recent years – and especially as our lives have become increasingly enmeshed with the Internet – the level of detail has increased, and our profiles have started to resemble the 3D character rendering from games that were made in the 1990s.

But if we want to capture a photo-realistic 3D picture of someone – along with an X-ray of their inner thoughts and dreams – we need to map them more closely from multiple angles. In the great ongoing game of data monopolisation, different watchers are trying to capture and hoard that data. Google has search patterns revealing your desires and intellectual interests, and Facebook has a treasure trove of its users' special moments, likes and projections of vanity. However, if we really want to see what a person *acts upon* in society – rather than their idle daydreams or aspirations – we should examine their payments data. How, where and when we spend is intimately revealing. This data gives any watcher deep insights into our priorities, habits and beliefs, and immediately shows what situation we are in. Briefly scanning through payments can reveal someone to be a well-off homeowner with an online gambling addiction, or a precarious renter who sends remittances to South Sudan.

Building up this data can in turn contribute to the ultimate prize – *predictive* systems that can accurately guess what you will do next, or how you'll respond to the next product or political provocation. For anyone looking to build such a character profile, cash is the great information blocker – it leaves no trails to be monitored and creates data black holes. If the payments sector can push cash users ever more into the digital realm, they will build a data goldmine of mammoth proportions.

An early record of disquiet about the power that financial institutions would accrue through this process can be traced back to 1968, when Paul Armer, a prominent computer scientist from the RAND Corporation, stood before a US Senate sub-committee to give testimony on the 'Privacy Aspects of the Cashless and Checkless Society'. He laid out his concerns in detail:

105

-6-

The extreme case, in which all transactions go through
the system and all the details are recorded (who, what,
where, when, and how) and then sent immediately to a single
center, obviously represents the greatest threat to privacy.
Such a system would know where we are and what financial
activities we are involved in everytime we so much as buy
a candy bar or pass through a toll station. Now it is unlikely that we will get to this extreme situation in the
near future, if ever. But how fast are we moving towards
it, even if we may never reach that limit?

Excerpt from Paul Armer's testimony

Armer was not the only one to be thinking ahead about the pit-
falls of a bank-dominated society. Malcolm Warner's 1970 book *The
Data Bank Society* was one of the first critical explorations of the
privacy implications of computer technology, and in it financial
data was flagged as one of the biggest areas of concern. In 1983, the
cryptographer David Chaum raised a similar concern about digital
payments systems, in a paper entitled 'Blind Signatures for Untraceable Payments'.

Chaum would go on to propose a new digital money system called
DigiCash to preserve its users' privacy. He is one of the technical founders of the 'cypherpunk' movement, which believed digital technology
was under threat by states and corporations that could use it for domination. Cypherpunks drew upon earlier anarchist philosophy, which
asserts that people, where possible, should manage themselves rather
than be managed.

Within anarchist political philosophy there is a power-balancing
formula that says *always support the underdog,* and, if an underdog
becomes a top dog, support the new underdog. In the realm of
information, this principle becomes 'transparency for the powerful, and privacy for the weak'. Historically, our world works the

BLIND SIGNATURES FOR UNTRACEABLE PAYMENTS

David Chaum

Department of Computer Science
University of California
Santa Barbara, CA

INTRODUCTION

Automation of the way we pay for goods and services is already
underway, as can be seen by the variety and growth of electronic
banking services available to consumers. The ultimate structure of
the new electronic payments system may have a substantial impact on
personal privacy as well as on the nature and extent of criminal use
of payments. Ideally a new payments system should address both of
these seemingly conflicting sets of concerns.

On the one hand, knowledge by a third party of the payee,
amount, and time of payment for every transaction made by an
individual can reveal a great deal about the individual's
whereabouts, associations and lifestyle. For example, consider
payments for such things as transportation, hotels, restaurants,
movies, theater, lectures, food, pharmaceuticals, alcohol, books,
periodicals, dues, religious and political contributions.

Excerpt from David Chaum's paper

other way around: large institutions cloak themselves in opaque
confidentiality laws and official secrets acts, but feel entitled to see
everyone else. It is like one-way glass: we cannot look in, but those
inside can look out. For example, a bank can demand to see my payment history when deciding whether to give me a loan, but I cannot
demand to see their loan history when deciding if I should accept it.

When I was a derivatives broker in the financial sector, my job
was to uncover information about which big bank was doing what,
which I gleaned through gossip and whispers from market participants. Later, once I had left the sector, I collaborated with civil
society groups on many different transparency projects, from
unveiling offshore companies used by large corporations, to helping Middle Eastern journalists understand company accounts. All
these projects involved investigating powerful organisations that

make themselves resistant to prying eyes. If you get involved in this work you realise that while dictators, business leaders and global corporate entities can use private Swiss bank accounts and chains of shell companies to avoid scrutiny, most ordinary people cannot set up these smokescreens. Instead, we depend on local financial data laws to eke out as much privacy as we can. Whether this works or not rests on the strength of our country's civil rights culture.

But as financial institutions push us towards digital apps, payments and online banking, we are ever more exposed to the potential of our data being used against us. In the age of big data and AI methodologies data has the interesting property of getting more useful the more someone has. Having a hundred data points can be more than ten times as valuable as having ten data points, and as the information accumulates so the temptation to use it increases.

So what types of payments data are sloshing around, and which watchers get to see it?

Primary Watchers: Financial institutions

Basic payments data includes who transfers what to whom and for how much, and where and when that happens. The institutions that get to see that data (and metadata that surround it) will vary depending on what path you take to initiate the payment. Your bank will see your every transaction, while your counterpart's bank will get a partial glimpse of you from the perspective of their own account holder. Card networks like Visa will see billions of transaction requests from hundreds of millions of people, while SWIFT sees huge numbers of international bank transfer orders. Groups like Google and Apple are creeping in via hardware devices that we may use to send the messages, while big intermediary platforms like WeChat, M-Pesa and PayPal will see all transactions initiated via their systems.

Financial institutions have historically been prevented from egregiously exposing our sensitive information to third parties. They are, however, permitted to augment their in-house data stores by forming reciprocal information clubs with other institutions. This is how many credit-scoring and fraud detection agencies work. Wolfie Christl of the Austrian research house Cracked Labs draws attention to how PayPal is entitled to share your data with 600 organisations, including banks, card networks, credit reference and fraud agencies, financial products providers, marketing and public relations companies, its internal group companies, commercial partners like eBay, legal agencies and financial regulators.

Financial data is used to categorise, label and rank people, and some of the most controversial players in data circles are data-brokers like Acxiom and Oracle. They have all manner of deals to gain access to purchase data, loan and income information, and credit card holder information from the likes of Visa and Mastercard. Google, for example, utilises sources like this, claiming that you can 'measure store sales by taking advantage of Google's third-party partnerships, which capture approximately 70 per cent of credit and debit card transactions in the United States'. This comes after Bloomberg reported that Google entered into a secret deal to purchase credit card data from Mastercard to track sales in stores. The search giant wants to sell advertising, and in that regard wishes to prove that customers who click on an ad will eventually purchase something. Credit card data is one way to establish this correlation. And in much the same way as Facebook builds an internal profile of you and then acts as a gatekeeper selling access to you to advertisers, banks like Wells Fargo and Citi build profiles of their customers' spending behaviour in order to sell marketing access to third-party product vendors.

Secondary watchers: States

Most states do not directly run the payments infrastructure, and thus are not conducting primary data collection. They can, however, request to dip into the data that banks and payments companies collect. One use for this is tax monitoring. A few years ago I found myself having dinner with a friend of a friend who turned out to be an official from the UK's tax authority, HMRC. She enthusiastically informed me that that they were actively working on proposals to accelerate digital payments. Two months later the UK Taylor Review called for a crackdown on the use of cash to pay tradespeople, citing the desirability of a digital record of transactions to monitor for tax irregularities.

A key element of all statecraft is measurement, as national statistics agencies and institutions like central banks gather data to assess policies and make macroeconomic projections. The thought of being able to gather micro-level transaction data on the entire economy is undoubtedly tantalising – though somewhat daunting – for state macro-economists. At an international level, the IMF openly makes use of the term 'surveillance' without it having negative connotations: 'A core responsibility of the IMF is to oversee the international monetary system and monitor the economic and financial policies of its 189 member countries, an activity known as surveillance.'

The words of the IMF do not raise alarm here because in the past these activities did not entail analysing the individual payments of billions of people. We are, though, moving towards a world in which the technical ability to do this vast-scale economic monitoring is taken as a given.

Perhaps most obviously, though, law enforcement agencies like to watch payments. I had a ringside seat when I was invited to be

110

part of the 2007 Cambridge Symposium on Economic Crime, at which operatives from the world's intelligence community arrived in droves, from Interpol and FBI agents to the Cayman Islands financial crime squad. The big topic at that time was the USA PATRIOT Act, the legislation passed after the 11 September terrorist attacks that widened the scope for financial surveillance by the US government. Section 314 (a) of the Act allows targeted individuals' accounts to be monitored, while various consortium efforts enable cross-border tracking of individuals. Banks are also permitted to share data with each other to track people who spread their money over multiple accounts.

If state security officials are after a particular person, they can find all sorts of ways to access their financial data. In 2010 an FBI document was leaked describing their 'hotwatch' system, set up to 'track the date, time and location of account transactions, as they occur'. The preferred method for setting up a 'hotwatch' is to sub-poena a card company like Visa or Mastercard – requesting real-time data on the target – while simultaneously issuing a court order to prevent them disclosing that they are doing so.

Things get weirder, though, when the authorities do not know *who* they are looking for, but do know *what behaviour* they are look-ing for. Authorities have long encouraged citizens to 'report suspicious behaviour', while authoritarian regimes push this to its most extreme form by building networks of fearful informants to rat on one another. But unlike the patriotic informant marching to the police to present evidence of a possible miscreant, banks see such monitoring and reporting as a costly exercise, not to mention a way to lose clients. They therefore have to be forced to report behaviour deemed suspicious. For example, in the US banks are required to file Suspicious Activity Reports (SARs) to FinCEN, the Financial Crimes Enforcement Network. In many countries finan-cial companies must scan all transactions and flag them to the authorities if they show certain patterns. This means customers' transaction histories are subject to analysis without a search warrant specifically identifying them as targets.

Despite the official narrative about cash being a channel for criminal payments, almost all law enforcement agencies have entire teams dedicated to tracking digital bank transfers for criminal and terrorist transactions. In 2017 I attended a security gathering at which an FBI director described how every day his team downloads the FinCEN database where all SARs are reported, and does keyword searches to sift through them before writing memos on cases to be investigated and pushing those to fifty-six field officers. Nevertheless, only around 10 per cent of the SARs get investigated and fewer still get prosecuted. Indeed, one of the major limitations of mass flagging systems is the ability of human officers to investigate the flagged cases. This is why automated surveillance systems – which we will meet again in a later chapter – are becoming the next frontier to conquer.

The systems described so far are all subject to due legal process, but payments networks are also subject to unauthorised infiltration by spies. Visa claims that it only hands over data when officially subpoenaed, rather than allowing open access to intelligence agencies, but that access may still be obtained. In the wake of the Snowden leaks, the German magazine *Der Spiegel* released an exclusive about the National Security Agency's 'Follow the Money' programme. Spiegel described how NSA operatives targeted entire regions, collecting some 180 million transaction records – 84 per cent of which were from credit cards – and uploading them to a private database called Tracfin, where they could be analysed. The SWIFT network is also subject to this spying. While some intelligence agencies like the British GCHQ have reportedly expressed misgivings about bulk data collection, these and comparable operations are very likely the tip of a surveillance iceberg.

The many faces of Big Brother

The term 'surveillance' inevitably evokes images of mass surveillance, but there are many more subtle variants of the practice. Most prominent is surveillance of people in a particular *category*. This could include, for example, an ethnic minority or a religious, political or socio-economic group. The scholar Nathalie Maréchal chronicles this in her academic paper 'First they came for the poor: Surveillance of Welfare Recipients as an Uncontested Practice.' As the title suggests, it tracks the history of surveillance of low-income benefit recipients. If you are an investment banker who keeps his huge salary as a result of a government bailout, your spending is not monitored, but if you are a former mine worker who got ill in poor conditions and now draws benefits you may be subject to checks on what you spend.

The paternalistic potential of payments surveillance is now being marketed to parents. It used to be the case that teens were able to escape the eyes of their overseers for moments of unmonitored exploration, but now parents can monitor not only their location via a smartphone, but also their spending via pocket money apps. Everyone can get a little piece of the surveillance pie, including yourself: budgeting apps and neo-banks like Monzo automatically categorise your spending into traceable histories. These technologies are sometimes said to be for self-tracking, but really it is the companies doing the tracking – and granting you some limited access to their results.

All these examples fall roughly into the 'Big Brother' category. They entail an authority figure – whether a company, government or parent – using payments data to monitor someone (or, alternatively, for you to monitor yourself). Merely monitoring, though, does not necessarily yield actionable insights. For example, if I am seen buying a briefcase and porcelain horse off eBay, while sending

113

three small payments from Bulgaria to Spain, what does it say about me? A human detective might treat those like intriguing clues from a Sherlock Holmes mystery, but the challenge for big finance and tech companies is to make sense of that without resorting to a human detective. This is why AI systems are being developed: rather than attempting to manually stitch together those clues into a narrative, an automated system may compare the pattern against vast collective datasets, matching it against similar recorded patterns in order to categorise you and produce automated predictions.

This means that data gathered through Big Brother processes can be used to produce two new archetypes. The first is Big Bouncer, represented by companies that use data to decide upon *access*. For example, automated credit-rating and fraud detection systems draw upon data to decide whether people seeking services from a firm will be let in. The second is Big Butler, represented by corporations that use data to profile customers they wish to approach. Presenting themselves as helpful servants, they use their intimate knowledge of your past behaviour to suggest things for you to do or screen out 'irrelevant' information. Big Butlers specialise in *steering*, or nudging behaviour.

We shall meet these Big Bouncers and Butlers again later, but they are proliferating around the world, and often in conjunction with Big Brother state systems. Despite paying lip-service to the classical liberal tradition – which places value upon privacy – US venture capitalists have poured resources into funding privacy-invading US tech giants to take over the world in the name of profit. At the same time the Chinese state has poured resources into supporting their own champion tech giants focused upon external expansion. Two of these companies are Alibaba and Tencent, and both have colossal payments divisions – Alipay (Ant Financial) and WeChat Pay. These mobile payments systems have made dramatic strides within China, and today are ubiquitous there. The WeChat app is like a multi-purpose Facebook with payments built in.

Western media have fixated upon the fact that even beggars in Shanghai carry QR codes to receive WeChat payments, illustrating how it reaches into the most marginalised corners of society.

But what happens if only two companies have a chokehold on the Chinese mobile payments space? Do they become huge honeypots of critical personal data? Well, yes. The Chinese central bank – the People's Bank of China – can gain access to the online payments data of people who use both platforms, and has set up the Nets Union Clearing Corp to get more oversight of mobile payments: normally mobile payments companies plug into the banking system in the background, but in this system they plug directly into the central bank.

This rise in dependence on mobile digital payments coincides with the early stages of the Chinese Social Credit System, which is an attempt to bring together a constellation of currently fragmented citizen monitoring systems (from credit scores to traffic offence registers) into a totalising whole. There are now many reports of Chinese people being blocked from travel and other 'privileges' if their score goes down.

Overt 'access control' systems like this are traditionally only found within formal organisations and corporations, which give privileges to those of higher rank (like the CEO being able to take the company jet, while the lowest employee must ride the bus). The idea of ranking, promotion and relegation systems being formally applied to people on a nation-state scale, however, is becoming technically feasible. While the Social Credit System looms as a prominent example, many of the technologies that enable this emerge from the activities of private-sector firms operating within a standard corporate capitalist setting. Not only does the global turn towards digital payment contribute highly sensitive data to help create these ranks, but it also provides states and companies with a vector via which to enforce restrictions. Freezing people out of a payments system becomes a means to discipline or constrain.

Payments censorship and panopticons

Freezing people out of digital payments systems – or limiting their access – is not a new phenomenon, but it stands to become a much more potent one as cash is marginalised ever further. One of the most well-known existing versions of this occurs in the realm of international payments, when banks make commercial decisions to block transactions from particular regions. With the rise of anti-money-laundering regulations, some banks have chosen to place a blanket ban on transactions to and from poorer regions. For example, British banks stopped processing Somalian payments, which severely restricted the transfer of remittances from the Somali diaspora in London. The major Somali remittance company Dahabshiil took Barclays to court in 2013 to protest at what it saw as the arbitrary shut-down of its Barclays account, which it required to facilitate cross-border transfers. It managed to win a temporary injunction against Barclays, but this threat of exclusion from the global payments system haunts many poorer countries.

While Barclays and other banks characterise this type of blocking as 'commercial', the use of banks for political blocking has traditionally been implemented through sanctions. Nation states can wield power by preventing their banks from processing transactions of key foreign businesses or individuals, or getting them to freeze accounts (such as those of former rulers like Egypt's Hosni Mubarak). In places like India, Greenpeace's bank accounts have been subject to freezes to stop it engaging in political activism, while Wikileaks was famously subject to a 'banking blockade'. In the latter, Visa, PayPal and Mastercard refused to process donations to the site, without being ordered to do so.

While the techniques described above are effective in the

international payment realm, and effective against wealthy individuals and institutions, they cannot easily be used against poorer people who operate with cash in more local settings. This is one reason why cash has a shielding effect for those on the lowest rungs, giving them breathing space. In a world where cash no longer exists, this changes. Two years after David Chaum released his paper, Margaret Atwood's masterpiece of dystopian fiction *The Handmaid's Tale* was published, set in a world in which women are subject to brutal subjugation. One of the key means by which this subjugation is enforced is the abolition of cash and its replacement by a system called Compubank, which enables the authorities to remotely monitor and control the women's lives:

I guess that's how they were able to do it, in the way they did, all at once, without anyone knowing beforehand. If there had still been portable money, it would have been more difficult.

Excerpt from The Handmaid's Tale

117

Imagine a theocratic state blocking the ability to purchase alcohol, or an authoritarian state punishing political opponents by placing limits on their spending. If you think that sounds unlikely, these systems are already being piloted: for example, the Australian 'cashless welfare card' stops welfare recipients spending in non-approved stores for non-approved goods. It is a perfect example of the forms of social control that can be exerted via digital payments. According to the *Economist* Intelligence Unit Democracy Index, over half the world's countries have governments that display authoritarian tendencies and disregard for the rule of law. Australian civil rights activists might be allowed to protest against the cashless welfare card, but such outcry will not be possible in many other countries.

When it comes to payments surveillance, people can feel torn between the prospect of slightly better safety against bogeyman terrorists, and the nervousness of what would happen if they got on the wrong side of dystopian payments censorship. The essence of the privacy debate, however, rests upon a more subtle point: imagine a situation in which two government agents arrive at your door, accompanied by two salespeople from a firm that provides remote video monitoring technology. The agents say, 'We have brought these men to install their cameras in your living room, bedroom and kitchen.' You say, 'But why?' They smile back and say, 'It's for your own safety, and if you have nothing to hide you have nothing to fear.'

This is a crude example, but it cuts to the heart of the matter. To a patriotic mind the argument offered at the door may superficially make sense, but we intuitively recoil at the situation regardless. And not because we have 'something to hide', but because it constitutes a massive overstretch into our private life, and treats adults as if they were small children. That is both infantilising and disrespectful.

The way that imaginary scenario plays out in real life, though, tends to involve the commercial players doing the talking, rather than government agents. Private digital payments systems expose their users to multiple forms of surveillance, but – when challenged about this – I have seen payments industry promoters use the same deceptive argument as above: they point out that they are good citizens with nothing to hide from authorities, and that you should be too. The implication is that a desire for privacy is tantamount to a desire to hide, rather than a desire to be treated as an adult. I can respect a legitimate state wishing to investigate me, but they must send agents to do the work. It is not up to me to do the work for them by volunteering myself for constant monitoring.

Unfortunately, technologies of micro-surveillance are proliferating across the board. I was confronted by one example of this while house-sitting a tiny studio apartment in San Francisco. After a day at the apartment I discovered that the owner had a small 'smart-home' camera hidden on the mantelpiece to remotely monitor her cat while she was away. I went pale. I had been wandering around her flat naked listening to loud music. Suddenly aware that everything I did in her studio could be watched, I immediately changed my behaviour, walking quietly around the room, fully clothed, and taking to crouching under the camera to get moments of reprieve.

This is what is known as the *panopticon effect*. As soon as you become aware that your private actions *could* be subject to secret monitoring, it dampens your spirit. The technical ability to monitor payments – regardless of whether the monitoring happens – has this potential, too. If things you buy can affect everything from your credit score to your insurance premium to your ranking as a good citizen, it will affect the psychology of your spending, and make you self-conscious. We currently do not experience this type of self-censorship very strongly, because our payments systems remain fragmented and split across physical and digital forms, but that is changing. A consolidation looms.

7

The Unnatural Progress of a 'Rapidly Changing World'

The term 'Stockholm syndrome' describes a disorder in which captives grow attached to their captors and then defend them. It was coined in 1973 after hostages locked in the vault for six days during a Swedish bank robbery later refused to testify against their hostage-takers.

Almost fifty years later, however, things have changed in Sweden. Many bank branches no longer have vaults, because over half will no longer deal in cash. Today, instead, the Swedish population finds itself bedding down in a new type of digital payments vault and growing ever more attached to the financial industry that watches over it.

But the Swedes are not the only ones getting drawn into corporate systems that offer superficially appealing services – we all are. It's wishful thinking to believe that this shift is primarily for our benefit rather than that of shareholders. In private, financial insiders tend to be sympathetic to my view, and occasionally send me emails admitting as much, but this issue now extends far beyond the boundaries of single countries and their local banking sectors. The digital payments vault is growing transnationally across borders.

In preceding chapters I cast doubt on the accuracy of the 'cashless' narrative, explored the class dynamics at play within it, and raised attention about the dangers of surveillance and censorship that accompany it. These alone should be enough to give a person pause about its desirability. But when I've been on shows laying out these arguments, the presenters often look at me as though I'm a fool trying to stop a river running downhill. 'Isn't it just inevitable that we will go cashless?' they ask.

Let's leave that question aside for now. What is more revealing is the assumption that underlies it. It is undeniable that we experience a *feeling* that the end of cash is inevitable, but where does this feeling come from?

Fake news futurism

Picture the aforementioned TV presenter redirecting the question to an industry futurist instead, asking 'Isn't it inevitable that we will go cashless, and if so, why?' Nine times out of ten, the answer will be something along these lines: 'Well, yes, consumers are turning towards digital payments because we all want convenience.'

Two features appear consistently in this style of answer. Firstly, *we* get credited for driving the change. Secondly, the inevitability is described as emerging from our desire for convenience. The assumption is that we are not content with the present, and seek out a future where ever more convenience awaits. So far, that seems straightforward.

In order to destabilise this assumption, let's consider its opposite. Imagine a world in which we have the option to stop, look around and be content with what already exists. Is it possible to promote such a world? Well, a person can certainly try, but they are going to run up against one formidable reality: it is antithetical to the structure of a capitalist economy, which relies on constant expansion in order to not break down. Economic growth has to be promoted at every opportunity, even to those who have more than enough stuff. In our world it is largely unspeakable to suggest some kind of 'stable-state' economy.

The requirement for our economic system to constantly expand is seldom overtly spelled out, possibly because it is so taken for granted. But it reveals itself through clichés, such as the perennial claim that we live in 'a rapidly changing world'. A favourite phrase of politicians and CEOs ('In this rapidly changing world, we must . . .'), it is used to spur people to either 'lead the way', 'keep up', or 'adapt'. The imagery is that of a treadmill that will dump us on the ground if we dare slow down.

To ask, therefore, whether people *desire* new digital payments systems largely misses the point. Maybe you do enjoy running fast on the economic treadmill, but those who do not will have to run regardless. Each individual person is a small node within a powerful system that transcends them, and – regardless of whether they enjoy the ride – the trajectory of corporate capitalism is not about to change simply because their desire to follow it is not present. We are tied into a structure that must expand, which in turn requires people to produce and consume more. This means anything that is slower, down-to-earth, non-automated or non-connected must be cast aside, like a restrictive skin that needs shedding. Cash, in this context, is a constraint on the further expansion and acceleration of our systems, preventing the treadmill from reaching ever higher speeds. What industry futurists will describe as 'consumers choosing digital payments', is actually people being *pulled* in a particular direction by the forces of global capitalism.

The glorification of inertia

In saying this, I am running counter to the traditional market economics narrative, which tends to claim that the world is driven by the active and informed choice processes of free individuals. I remember facing off against a German economist who scoffed at my suggestion that people are pulled into digital payment by forces beyond their control. 'Who are you to say what drives people's choices?' he flashed angrily. 'Just let the market decide!'

My response was to ask him why he had *decided* to wear a suit. He was taken aback, because only a tiny percentage of people who put on suits to go to work make a conscious choice to wear them. Most suit wearers conform to the cultural convention out of habit and fear of being judged by their networks. They are implicitly *told what they want* by their peer group, and turn up to a store to buy one. The only personal decision concerns surface-level customisations like what colour the suit will be, or its patterning. This is how offices end up full of people who have decided to buy suits, but to describe this inertia as 'choice' is highly uninformative, if not downright misleading.

Anthropology is a discipline dedicated to studying these collective cultural force fields that people find themselves within. Mainstream economics, by contrast, has a bad habit of implying that we are autonomous individuals making choices in a free-form environment. However, not only are we caught up in cultures, but those cultures themselves are also set within much larger economic networks that pull on us like a mesh pulls on the individual nodes that constitute it. The largest nodes in this mesh gather together into powerful oligopolies – small groups of huge players – that have far greater power than us to establish precedents and set changes in motion.

If a market is dominated by such an oligopoly, then letting 'the market decide' is not that far off letting the oligopoly decide. Oligopolies are like menu-setters who place certain options in a good light with large font. Consider the fact that Apple is supposedly a private company, and yet finds its actions reported on the front pages of newspapers (as if they were crucial political events), giving it priceless free advertising and automatic market dominance. There is nothing about using the Apple Pay app that suggests small consumers are shaping the future. It is quite the opposite, and yet it will be reported on as 'Consumers opt for mobile payments,' with no mention of the menu-setters.

Stories about 'consumers' therefore need to be taken with a pinch of salt. In the same way as someone caught in rapids may have five possible paths to choose from between the rocks, we may have options within an otherwise non-negotiable direction of travel. For example, will you choose Apple Pay, Google Pay, contactless card from bank, mobile payment app, or some other type of cloudmoney? Cash is not on this menu, even though it's used by many people, because it doesn't support economic acceleration, and is not within the accepted direction of travel. That direction is set in motion by the big players who stand to gain the most from it within the economic network.

Thus, while many economics commentators focus on the *active choice* element of markets, it is more revealing to focus on the *passive* element, in which people are just pulled along. A useful metaphor to convey the ambiguities of this is that of *slumping*. Maintaining good physical posture requires an active choice, whereas slumping is a passive process in which I unthinkingly let gravity take over my body. The feeling a digital payments user gets in places like Stockholm is far more like the latter than the former. The 'gravitational pull' of digital payment is rapidly increasing as ever greater numbers of people lose the willpower to avoid sinking into financial institutions. All the invisible chains of dependency that exist within the interdependent economic mesh are pulling on people and warping the structure of their economy. Because that new situation is getting baked in as the default in all systems – for example, parking meters suddenly 'go cashless' – trying to use cash in that situation will very much feel *active*, like trying to maintain posture: the cash user increasingly feels out of sync, as if resisting a tide. That gets strenuous, so it is easier to just let go and sync up. The latter is a *passive* process, but one that will be reported on as an active consumer choice.

Catalysing a slump

This dance of active and passive is a repetitive cycle in capitalism, but the constant requirement to sync up is not welcomed by everyone. Grandparents, for example, are often unimpressed with products their teenage grandchildren are obsessed by, because they have already been subject to wave upon wave of industries telling them what new things they should desire. When they were teenagers they too were probably anxious to keep in sync, and by the 1970s would have been cast as the stars of a cutting-edge economy by being compliant consumers of TVs and microwaves. After a few decades of this, the message of endless product change begins to feel empty. Once this feeling sets in they are then implicitly told to 'make way for the future'. To modern digital giants the elderly shopkeeper in a small English town, or the traditional fabric merchant in Mumbai, is just human friction standing in the way of the large-scale profits that will accrue from large-scale automation.

Many of these people are still an important part of the economy, though, and their resistance becomes a block to the expansionary

126

tendency of the system. They are told that they should want ever more speed and digitisation, but humans are slow, tactile beings, and one major reason people cite for preferring cash is its tangible physicality, the visible and textured tokens helping them to intuitively budget. Several studies have shown that digital payments encourage fast and detached spending that more easily leads to indebtedness, but that is part of the point: in its 'benefits of going cashless' website, Visa reports that 'a US consumer typically spends 25 per cent more on a card at a pizza shop, 33 per cent more at a deli or diner, and 40 per cent more at a family restaurant'. The terms accompanying digital payment ('contactless', 'frictionless') promote a state in which you rush through life consuming things, which is required within an expanding system that has no time for your moments of pause.

Understandably, many people are stubbornly resistant towards entering this state of being, but – as suggested in the previous section – non-compliant people can eventually be dislodged, provided their environment is altered to such an extent that they have no choice. This can be speeded up by carefully planned interventions. In 2019 I met Badal Malick of the Cashless Catalyst initiative, a USAID-funded program to convert people in India to digital payment. Badal described the 'co-ordination problem' he faces: to get Indians to lock into digital payment requires getting enough merchants to accept it and enough customers to use it. Catalyst thus experiments with *catalysing* mechanisms, such as getting big companies to pay their staff in digital money, or offering digital credit to small merchants to 'hook' them into digital payment. These procedures come out of research from USAID's Global Development Lab, which has four official goals: advancing US national security, tackling the root causes of immigration and terrorism, creating economic opportunities for Americans and, finally, 'Supporting US business interests', to which end it partners with – among others – Visa and Google. It has a manual about how to digitise the Indian monetary system, and now they fund Catalyst.

Badal shared all this with me in London in 2019, a year in which the results he sought in India were rapidly unfolding among London's middle classes, who were slumping into digital payments in a self-reinforcing loop while bougie 'cashless' shops blossomed. Once one part of an interdependent network begins to slump, it can be used to pull in the rest. The hipsters are currently being used to make cash-heavy communities feel as though they are out of sync. The cashless coffee shops popping up in my old neighbourhood of Camberwell in South London stand out in stark opposition to the cash-only Ghanaian hair salon or Eritrean general stores on the same street. The rise of digital payments runs parallel to the process of gentrification, in which 'rough' shops with an informal ethos are displaced by boutiques that will pave the way for standardised chains.

The gentrification of payments

Once you view the global economy as a transnational network, it becomes possible to see how trends set in global cities in powerful countries can induce payments changes within tiny rural towns in poorer countries. To someone growing up in a marginal country like South Africa, London and New York loom like mythical kingdoms of sophistication. Almost every middle-class South African feels drawn at some point towards making a pilgrimage to see these top-tier cities. The world's airports are crammed with such people from all over the globe. The aspirational South African sits in New York's 'cashless' La Guardia Airport, before boarding a 'cashless' British Airways flight to London, where the bus system refuses cash, before completing the tour in Amsterdam, where the tram system refuses cash. They are left with a clear message: *cashlessness is endorsed by countries much more internationally important than your own.*

It is from the ranks of this transnational middle class that many local entrepreneurs in any country will emerge. They take on the trends they are exposed to on their travels, and this creates cultural bridges between cities like New York and, for example, Johannesburg or Nairobi. These cities in turn influence smaller towns in their own countries, and digital payment spreads alongside these bridges to make incursions into cash economies. For example, the tiny Andean town of Pisac in Peru is home to indigenous Quechuan women who sit in the square selling vegetables to locals using cash, and yet when I entered the town in 2018 I was greeted by a giant Visa billboard. That's because Pisac is also a mecca for bohemian international travellers who like to perch in tourist balcony cafés set up by local jet-setters, using the Wi-Fi and their cards as they look over the old informal market.

Each of these new businesses lays one more strand in the digital payments web being woven. This has knock-on effects for far humbler merchants who target a different demographic. Pucallpa, for example, is a dusty town in the Peruvian Amazon region, in what used to be primary rainforest until the logging, mining and palm oil plantations spread. The low-paid workers here use cash, but the town has also become a minor centre for the growing ayahuasca tourism trade, with foreigners arriving to drink the psychedelic sacrament. For some in the indigenous Shipibo-Conibo community this development is amusing, and it also creates new opportunities to trade. But along with their shamanic enthusiasm these foreigners bring cards (or Apple Pay). While I was there, I entered a rickety store as a shopkeeper squinted through a set of faxed instructions telling him how to set up a new point-of-sale card terminal. It lay beside the small plastic dishes on his counter, which he used to keep coins to give out as change to his local customers. He was tethering himself into the global cloudmoney system.

This man will be praised by official institutions – he is adapting to keep up with a rapidly changing world. If you look at all the POS terminals and fintech apps in the context of the global economy, they are like front line agents in a process of global gentrification, paving the way for the card companies, Big Finance and Big Tech to overrun informal economies. Gentrification is just the leading edge of *corporate seep*, the process by which previously informal and direct peer-to-peer economic relationships are replaced with institutionally mediated ones.

It is in this context that 'financial inclusion' needs to be analysed. I have already suggested that – in mainstream circles at least – inclusion is more akin to *absorption*. A small, informal 'Chama' savings group managed by elderly Kenyan women is seen as quaint, but inferior to a large-scale app presented at TechCrunch Disrupt by flashy entrepreneurs. Financial inclusion is not envisaged as thousands of Kenyan women's co-operatives flourishing. Rather, it entails dissolving those informal systems, and onboarding the women into the large-scale corporate systems that will replace their co-operatives. Doing an image search for 'financial inclusion Africa' is a rough-and-ready way to confirm this: it brings up pictures from Mastercard's website, all of rural women smiling into the screens of their mobile phones, presumably looking at an app tethered to a distant data centre (run by people who look nothing like them).

Not all financial inclusion professionals are supportive of this corporate seep, but many are aware that – in a world increasingly dominated by those powers – people who are not absorbed under them could end up discriminated against. We're back to Stockholm Syndrome. When their dominance is so apparent, you want to see the good in your captors, rather than the bad. It's easier to take refuge in the idea that they are benevolent providers of consumer benefit, and – with this mindset in place – both centrist liberals and mainstream conservatives can agree that when the card companies reach rural Zambia, the people there are 'catching up' with progress.

This ideology appears everywhere. It is why Western corporates like Gap, Unilever and Coca-Cola have all signed up to the Better than Cash Alliance, pledging to pay their outsourced workers in poorer countries with digital payment. It is why major humanitarian NGOs like CARE, Mercy Corps and the Save the Children have signed up too. Outside an Athens hotel I bummed a cigarette from a man from the Catholic charity Caritas. He was at a gathering to discuss how to digitise aid money previously handed to needy people (in disaster zones or war zones) in the form of cash. Outside that same hotel a man from the aid community spoke to me in conspiratorial tones as he showed me a Mastercard-branded card, issued by the Turkish Halkbank on behalf of the Turkish Red Crescent, which then handed them to refugees. He claimed that this three-way partnership brought the bank a huge number of new customers and suggested that I should look into the deal as an investigative journalist. But these kinds of deal – in which Mastercard partners with Mercycorp, the World Food Programme, and numerous governments alongside banks – are now so prevalent that they are barely seen as noteworthy.

Indeed, the most common mainstream 'critique' of this move towards a vortex of digitised payments is that it has *not yet engulfed everyone.* I appeared alongside a Citigroup executive on a media discussion, and when he was asked what difficulties could emerge from a cashless society, he said 'ensuring full inclusivity'. In his mind the main problem is that not everyone has been on-boarded to his industry's systems. He concluded, 'It will be a shared responsibility between the banking industry and governments to ensure this happens.' Translation: the state must help us sign up new customers.

This corporate message is often inadvertently reinforced by well-meaning but misguided social workers, who talk about how we should be sensitive to the fact that some people are not ready for the transition to the digital economy. They argue that people who are slow to *adapt* need to be saved for fear that they may be *left behind.* They are presented as having missed a ride, rather than as having rejected the direction of travel. Thus, in the UK, a glut of news stories emerged in 2019 about homeless people, people without bank accounts and elderly people (who may have bank accounts but who prefer cash) who could be stranded as cash is throttled and ATMs and branches get shut down. These stories are followed up by heart-warming rescue stories, in which government, charities and 'tech-for-good' start-ups do things like hand out card readers to homeless people. The interests of the corporate sector are being presented as charity.

It is easier to go along with corporate seep than it is to resist it, but some local governments are nevertheless trying. For example, the municipalities of Philadelphia and San Francisco introduced laws in 2019 requiring shops to accept cash. Within the ideology of endless growth this type of action is easily presented as one of those futile 'Luddite' acts. Tellingly, however, stories surfaced about Amazon doing behind-the-scenes lobbying against this legislation. The company complained that the legislation would stand in the way of its fully-automated Go Cashierless stores. The *Wall Street Journal* questioned whether the legislation placed 'limits on innovation', but it is far more accurate to read that as 'limits on automation'. So far I have presented the drive against cash as being beneficial to financial institutions, but they are in a symbiotic relationship with large digital corporates: to Amazon, my fumbling around in my wallet for a cash token feels like an eternity of uncertainty and incompatibility that jars with its system. A cash payment cannot be remotely initiated, monitored, co-ordinated or stopped, whereas the digital money infrastructure resonates in harmony with its ethos. Cash is a bug, jamming the emerging fusion between finance and tech, and given that those are the biggest players in our economic network, they are jointly pulling away from it, and pulling many people along with them. That is why it is being crushed, or, to put it in its more traditional framing, that is why *consumers are turning to digital payments in this rapidly changing world*. Let's look more closely at the emergent fusion.

8

Shedding and Re-skinning

Throughout the ages, bankers have accumulated money by writing and trading contracts about money – and throughout the ages this has been a source of moral panic. A Florentine Medici banker in 1416 needed only capital to back contracts or absorb losses, and a quill pen to write the contracts. It did not take much physical energy, but it took mental energy to think everything through, and emotional energy to stomach the risk of loss. For a person used to physical labouring in those times, such activity might seem mysterious – even devilish.

But, no matter how devilish they might have seemed, early bankers were visible members of a community. The Japanese moneylenders in medieval Kyoto would situate themselves in *sake* brewing houses, while the seventeenth-century financiers of London's Square Mile lounged in coffee shops. Over time, as these localised financiers joined forces to form bigger syndicates, they might establish a series of out-posts in different places – a *branch network*.

The banking giant Barclays PLC, for example, started from a seventeenth-century partnership of Quakers, and grew in the nine-teenth century through the amalgamation of several smaller Quaker regional banks. As the bank centralised towards its modern corporate form, its branches nevertheless maintained aspects of decentralised localisation: a farming town in the 1880s might have a branch overseen by a branch manager with localised power, who reported to the headquarters in London (which would oversee the wider branch network while acting as a central point to consolidate risk and raise financing). Even if the headquarters were out of your view, you would still be able to shake the hand of the local man who assessed you and handed you the contract to sign. Increasingly, though, we cannot see financiers, or shake their hands. Rather, the only thing your hand may end up touching is the interface of a digital finance app. In this chapter I will show you this trajectory, and how it blends into the parallel trajectory of Big Tech.

The federated front line

Many large organisations have historically operated through a centralised–decentralised model in which the central headquarters posts representatives to front lines: the state sheriff was posted to a frontier town, an ambassador to an embassy, and a missionary to a distant rural hamlet. These representatives extend the power of the central organisation, but simultaneously introduce deviations from its core mission. They could go rogue, be corrupted or become saints. Similarly, the local bank branch manager could be an incompetent and petty racist, or a cold but fair captain, or an empathetic and loved member of the local community. Thus, the same central entity is transformed at its edges into a mottled 'federated frontline'.

But as twentieth-century suits replaced nineteenth-century top hats, technologies of bureaucracy – like faster postal and telecommunications systems – could tie these customer-facing branches in more tautly to the central corporate headquarters. Rather than deferring to the local knowledge of branch managers, banking corporations could hire strategists at their headquarters to study data from the front lines and develop formal guidelines for decision-making. The federated front line began to look more uniform as its staff slowly had their autonomy eroded and were subject to greater levels of standardisation. If branches originally were like chefs drawing from a central pantry, they became more like waiters mediating between us and a central chef.

As they standardise, large-scale institutions lose the desire or ability to adapt to the circumstances of each person that approaches them. Rather, the customer must do the adaptation as they are presented with standard menus and rule sets: *If you want W, fill out form X and present it to counter Y who will send it to headquarter department Z.* The only thing that stops the customer feeling alienated is the branch clerk, who might guide them through the application with a smile and a friendly word.

Under the capitalist imperative to cut costs and expand, firms must not only standardise processes, but also seek to automate them. We have become increasingly resigned to this idea, but in the 1950s this creeping sense of technological bureaucratisation was unnerving to many. In the July 1958 issue of *New Scientist*, Barclays Bank felt compelled to run an advert called 'No Robots at the Counter.'

NO ROBOTS AT THE COUNTER

At Barclays we pride ourselves upon a human understanding of our customers' needs and financial problems. Science has given our staffs a wide variety of mechanical aids, ranging from simple postal franking machines to the very latest in electronic calculators and computers.
But these do not obtrude. They are simply a guarantee of speedy accuracy which lies behind a friendly and courteous service.

BARCLAYS BANK
Money is our business

Sixty years later, however, banks are going all out to build non-human 'counters' to bypass branches and plug us directly into their core systems. If they were to rewrite the advert now, it might read like this:

NO HUMANS AT THE COUNTER

At Barclays we pride ourselves upon a technological understanding of our customers' needs and financial problems. Science has given the company a wide variety of mechanical aids, ranging from machine learning to facial recognition technology. Our staff do not obtrude. They are simply there to maintain the machines that are behind your self-service apps.

Re-skinning big finance

Branch counters were the original 'user interface' for banks, the connection point between a user and the system they use. In a modern bureaucratic world we are surrounded by user interfaces that plug us into distant technical systems that are too confusing for us to deal with directly. Much like a plug socket conceals a vast fossil-fuel electricity complex, interfaces conceal what is going on behind the scenes, to the extent that we never have to understand it, or even know it exists.

For example, behind the scenes of the banking sector there exists an interbank market that allows banks to do private deals with each other. This is invisible to members of the public, who might 'plug in' by walking into a branch where a clerk shows them a glossy brochure for a fixed-rate mortgage. These customer-facing service staff provide a buffer zone, shielding the technical people and senior management in the core of the bank while funnelling information to them, a process which in turn will express itself in private deals in the inter-bank markets.

These clerks, however, introduce unpredictability. They can spend too long chatting to customers, can antagonise people, rebel against their bosses, make mistakes, go on lunch breaks or slow things down with their attempts to be humane and empathetic. They also need to be paid. Banks inevitably wish that this branch interface could be shed and replaced with a more direct line. Their initial attempts at this included phone banking – getting people to call them rather than using branches – and payments cards, which are ways to order bank transfers without directly visiting them. What was truly needed, though, was a way to make a bank branch materialise in someone's home.

This breakthrough came from Silicon Valley. Big Tech has a behind-the-scenes army of engineers building its core systems, but these companies distinguished themselves by porting those systems directly into people's *private space* via apps hosted in their hardware devices – originally a home personal computer. As long as the telecommunication network stands up, the illusion of a 'store front' can be constructed on your home computer.

Banks were initially slow to co-opt this technology for their own purposes. In the 1980s they were experimenting with what they called 'home banking', which by the late 1990s became known as online or Internet banking. By the 2000s, Silicon Valley was converting its porting technology to smartphones, devices that live in your pocket. In the age of smartphones, the phantom 'store front' sticks to you as you walk around.

Global banking giants, however, have multiple priorities, and can only devote a certain amount of their attention to this question of how to erode their branch networks. During their meetings the management also have to consider other things, like how to manage multi-billion-dollar portfolios of mortgages in an uncertain environment, or how to deal with the geopolitical risks of financing Russian gas exploration. They are, to put it mildly, juggling many things at once, and have millions of legacy customers. This means they move comparatively slowly.

This sluggishness allowed a moment for a new class of entrepreneur to front-run the banks. While banks had been working with technology for decades, a new buzzword – 'fintech' – rose to prominence in the 2000s. It referred to any company that sought to import Silicon Valley-style interfaces into finance, but without being encumbered with any legacy business. Fintech teams were small and specialised, and their meetings did not have items like financing Russian gas exploration on the agenda. This is because they were not financiers. They were builders of digital 'store fronts'.

For example, the Level 39 fintech accelerator (with which we began Chapter 1) hosted Revolut, a start-up playing on the word 'revolution'. In reality, however, the original Revolut app was just a skin pasted over other financial companies – a third-party storefront that would take your business and pass it on to background financial institutions.

But in the aftermath of the 2008 financial crisis, these app-led systems were increasingly pitched as an 'attack' on the old banks, as fintech firms sought to associate the imposing physicality of grandiose bank branches with outdated sluggishness and bloated corruption. Fintech start-ups could – initially at least – take on the heroic persona of 'disruptors'. They pitched themselves as champions of the youth, calling for 'the new'. As with digital payments, the narrative settled on the story that digitisation is driven by 'changing customer expectations', with millennials and Generation Z often cited as 'demanding' these digital finance store fronts.

The media largely accepted this story, and – as with digital payment – there was little critical reflection on how expectations are engineered. Generation Z came of age as awkward teenagers in a world already dominated by tech firms, and had no say in the establishment of those omnipresent behemoths. Given that Silicon Valley's systems were in the process of warping the structure of the economy – and accelerating it – fintechs could claim that the lack of apps offering instantaneous access to finance was an injustice. Their 'revolution' was to bring the gospel of appification into any element of finance not yet fully appified, be that payments, lending, stock trading, investing or wealth management.

This battle cry initially struck a nerve, as banks felt some pressure – real or imaginary – to catch these Silicon Valley aesthetics up. Pundits took advantage of this brief moment of industry insecurity to predict the end of banks. Almost every tech conference of the past decade has included a panel entitled 'Will tech take over finance?' or 'Do we still need banks?' Not only were these fintechs supposed to disrupt finance, but they were also supposed to *democratise* it: fintech would deliver financial access to a wider user base while breaking down the power of the old ruling class.

The pseudo-revolution

It has always been a dubious story, though. Consider any small app on an iPhone – for example, a gym training app – and ask yourself, 'Does this disrupt or strengthen Apple?' Of course, the myriad of small independent apps built upon Apple's IOS operating system strengthen Apple, because in using the small app we end up using their operating system.

Similarly, in using small fintech companies we often end up using the same underlying financial 'operating system' that is controlled by existing powers. This is because fintechs, on average, do not *bypass* the existing financial system but plug into it. They use the ecosystem collectively controlled by major banks and payments firms like Visa as a substrate upon which to build their budgeting tools, savings platforms and so on. PayPal was one of the original examples of this – it started with big claims about disrupting finance, but in the end proved to simply be a 'plug-in' money transmitter attached to bank accounts.

Just as Apple is not going to deploy its engineers to build gym training apps, but is happy to serve as an environment to host them, so too are banking oligopolies happy to host accounts for niche fintechs. Rather than a bank trying to deal with the admin of tens of thousands of micro-savers, it can give one account to a fintech who aggregates those, automating the interactions with an app to make them more profitable before plugging into the underlying core 'operating system'.

This 'operating system' is underpinned – in the final analysis – by states. If fintech is based on automating access to finance, and finance is about contracts for money, and the monetary system is underpinned by national banking oligopolies, it means fintech companies often need to *partner* with banks to have any ability to operate at all. This is why many digital 'neobanks' – which in Europe include players like Monzo, N26, Revolut and Fidor – were originally just interfaces pasted over the old banks, reflected in the fact that they originally did not possess banking licences. Traditional banks start as actual banks and then later develop interfaces to engage the public, but neobanks *start* as interfaces.

Many of these fintechs have about as much chance of disrupting finance as an IOS developer has of disrupting Apple, and they cannot wrench the monetary system away from banks any more than an app company can take control of the iPhone. The only ambiguity that emerges is when fintech companies cut directly into something banks would otherwise do, but after years of fintech rhetoric we have seen, overall, little bank disruption. The apparent death fight between banks and fintech looks more like symbiotic partnership that entrenches the overall power of Big Finance by extending its reach via new players.

The delusion of a revolutionary uprising can persist for a long time, however, because the story serves both as a catchy marketing line for fintechs and as a cover for banks, who can justify their own desire for automation as a 'fight for survival against fintechs'. The banks always wanted to automate anyway, but had a bigger 'to-do' list, so were slower. They are still here, big as ever, actively absorbing fintech technology into their own systems by buying them up. They let the fintech sector run the risky experiments – funded by the founders and VCs – and if they are successful, buy them, copy them or offer to run the background plumbing they require for their platforms. Ten minutes' walk from its colossal tower in Frankfurt, Commerzbank has set up its own fintech incubator – called Main Incubator – to be the 'laboratory of tomorrow's bank'.

The fintech industry operates in an almost permanent state of doublethink, which is why a peculiar atmosphere hangs in the air at Level 39. The *aesthetics* of fintech accelerators are different to those of traditional banks – with hackathons on the future of money and pitching contests full of the language of disruption – and yet there is a pervasive feeling of people play-acting. After all, the venture capitalists will make sure the start-up they are backing will work together with the big banks – or sell out to them – if that is what it takes to make returns.

Over the years both banks and the fintech industry have slowly and quietly acknowledged their symbiosis. After thousands of industry articles and pundit predictions about the imminent destruction of banks, conciliatory pieces about how banks and fintechs could 'work together' emerged. These articles are always presented as a kind of 'a-ha' moment, when the industry thought-leader or analyst realises that the encounter they've been reporting on as a battle was actually an uncomfortable series of first dates between future spouses.

Direct line

In 1987, when I was a little boy, my artist mother took me to an exhibition of an up-and-coming illustrator called William Kentridge. His emotive sketches are the first artworks I remember seeing, and were being shown at South Africa's National Arts Festival. The festival was sponsored by South Africa's Standard Bank, and Kentridge had just won the Standard Bank Young Artist Award. The artist and campaigner Mel Evans refers to this phenomenon by which coldly commercial institutions patronise art as 'artwash', but banks are no longer simply associating themselves with artists. They are *hiring* them to make their new digital interfaces attractive.

In 2019 I got access to another display of Kentridge's work, but this time it was projected onto a screen in a seminar space on the seventeenth floor of the headquarters of Lloyds, the UK's second-largest bank. A manager within the bank's design team was streaming a Kentridge video animation called 'Second Hand Reading', in which contemplative figures walk against the backdrop of an *Oxford English Dictionary* to an African jazz soundtrack. He was giving a heartfelt lecture on how the piece exemplified the interplay between words and images, a fusion that was important for the bank, which needs to create elegant text and images for the corporation's fintech apps.

The team members he was addressing were young and had a background in art and design. Their job was to help with the bank's 'digital transformation' – the process of shutting down branches and pushing people onto apps. Banks might have previously hired interior designers to deck out branches, but the job of drawing people into apps requires *user experience design* (the digital equivalent of aisle layout). The team's big boss is the Chief Design Officer, a former head at Google brought in precisely because of his expertise in digital self-service systems. The mission is simple: move people away from costly service and get them to give themselves cheap self-service.

Earlier I noted that bank branches are like waiters mediating between us and a central chef. Truly good waiters make an art form out of service, which entails matching or adapting the capabilities of a chef to the requirements or desires of the guests, perhaps allowing idiosyncratic alterations to the menu or giving suggestions of their own. 'Self-service', on the other hand, is the touch screen at McDonald's that replaces that waiter with hardcoded options to be communicated directly to the kitchen. Insensitivity to customers is not just a by-product of this – it's the point.

Corporations justify self-service by telling you how fast it is, but in the long term replacing service staff with machines allows corporate management not only to cut costs, but also to standardise and gain direct control over customer options and experience. Self-service apps also allow them to extract far greater amounts of data about you. Thus, while some people may like apps and others not, the banks are going to push you onto them regardless, because they have a commercial imperative to do so. And in the same way as they have to change attitudes towards cash, so they have to find ways to wean non-compliant people off physical service. I have worked with user experience design teams in London who are given explicit briefs by banks to help them solve the 'problem' of older people who still expect this service.

Financial insiders use all manner of euphemisms to talk about these non-compliant people: a statement like 'There are obstacles to digital transformation' translates into 'We are struggling to push people onto our app.' The decision has been made, though, and we already saw how this formula plays out in the previous chapter: the big players decide on the direction of travel. For example, in 2019 Nationwide ran the 'Here Today. Here Tomorrow' billboard campaign in the UK. It came in response to newspaper articles warning about small towns being stranded without bank branches as banks closed them down. 'We promise,' said the billboards, 'that every town and city with a Nationwide branch today will still have one for at least the next two years.' The translation: *You have two years to transition to our app.*

Once shepherded onto apps, however, people can feel confused and alienated. Lloyds has over 20 million customers, so any minor errors or confusing layouts in its self-service interfaces have the ability to send millions of people down the wrong track. A basic interface just needs a list of options and a way to choose them, perhaps alongside an FAQ to answer common queries, but it can be an isolating experience. Nowadays we find self-service machines that attempt to speak to us – like the self-checkout machines at supermarkets robotically saying 'Place item in the bagging area' – but the experience feels empty because it is just the delivery of non-negotiable instructions.

One way to dress up this mechanical experience as a human one is to programme self-service systems to use pronouns like 'I', 'me' and 'my' ('Place the item in *my* bagging area'). An even more advanced technique is to add in the illusion of two-way interaction. This is where Natural Language Processing (NLP) comes into play: NLP converts human speech uttered by you (or words written by you) into a language the computer can understand, allowing it to guess what you might be trying to ask, a bit like a dynamic FAQ directory that can rearrange itself. When this is combined with the new practice of using first-person pronouns, the 'chatbot' is born, a digital interface presented as a living being.

Financial institutions are now obsessed with such chatbots. The first I came across was Cleo, which lives within the confines of an iPhone app. I met 'her' at a fintech pitching contest where 'she' was being demonstrated to venture capitalists, who could watch the system answering queries.

Hey Cleo, what's my balance?

Hey Alex! Mastercard: -£760. Current account: £1048. Savings: £1700

Cool, how much have I spent at Pret this month?

You've spent £44 at Pret since you got paid on 15 December

This 'AI assistant for your money' is not the first bot to come with a female name (there's the scheduling assistant Amy, and Amazon's Alexa), but these systems all try to get on first-name terms with you. Given that a bot is an interactive interface for a company, its pronouns must refer to that company. This is curious, because while we might have once accepted a statement like 'Would you like *us here* at Bank of America to give you a mortgage?', we would have found it absurd if Bank of America pretended to be a *real* person by saying 'Would you like *me* to give you a mortgage?' Bank of America, though, can now do this, because it has a digital interface with a human first name – Erica.

These shapeshifting digital interfaces with names are an evolution in *corporate personhood*: financial institutions are re-skinning themselves in a machine shell that no longer even refers to their employees, but to *itself*. This quest to give automation a personality goes beyond the visual: HSBC, for example, is branding the 'sound of HSBC' for its chatbots.

The next phase of this shapeshifting is to add personalisation, with the interface morphing to mimic your accent, turns of phrase or favourite emojis, like a chameleon skin that changes depending on who touches it. Historically, branch managers could use their memory of your past interactions with them to inform new interactions; likewise, these automated systems can log your behaviour to inform the way they subsequently address you. The key difference, however, is that the federated front line is gone. In the past, you might fail to get a loan from the branch manager in one town, but convince the manager in another to trust you. As banks rip away that patchy human interface and replace it with a uniform digital one (however superficially personalised), that diversity becomes uniformity. This effect is enhanced when separate banks begin to share background systems, such as common credit scoring systems.

This re-skinning is occurring on all fronts. One of the biggest areas for banks to automate is customer helplines. Customers already hate the robotic way that humans in call centres are forced to interact, so it comes as little surprise that big institutions are happy to invest in robotic systems mimicking humans. The problem, though, is that while we may go through the motions of interacting with some chatbot mimicking a human in low-stress scenarios, in situations of high emotion (such as calling up the bank to inform them that your parent, who is their customer, has died) the chatbot's fake emotion rings hollow.

Thus, while banks all agree they will shed their branches in favour of a digital shell, their managers express occasional concern about this breakdown in human contact. They pledge to maintain some 'points of presence', by which they mean actual employees capable of empathising. In the grand scheme of things, though, these small remnants cut against the de facto trajectory of corporate capitalism, which is not encumbered with sentimentality, and seeks out only efficiency, speed and scale.

The fusion

But if the banking sector has imported the shape-shifting techniques of Silicon Valley (via the fintech sector), then it has given Silicon Valley new inroads into finance in the process. By building out their digital capabilities, banks make their systems increasingly compatible with those of Big Tech, because both revolve around the same concept: core systems that are ported via apps to millions of

148

customers with accounts, who in turn use those apps to send billions of messages to the institutions' data centres. It makes sense for those institutions to link their data centres together to develop synergies.

For example, if Uber's 'brain' is its data centres, its 'thought' involves processing those millions of smartphone communications it receives. But this does not translate into human action unless paired with a payments system. Uber drivers are not going to take orders from the Uber brain unless they believe they will be paid. Thus, Uber pairs itself with financial institutions, which now begin to resemble a 'motor cortex', the part of the brain that translates thought into action. It is an obvious step to merge these elements into one, which is why almost every major tech company is entering into partnerships with financial institutions.

Given that people are tethered to both Big Finance and Big Tech by accounts, linking these accounts becomes the initial means by which institutions fuse: I tie my Amazon account to my bank account, giving them permission to become partners. The old days of walking anonymously into a frontier store with paper dollars is transformed into a store front materialising on my screen, with a single click triggering both a bank transfer and delivery of goods. This is why Amazon lobbies against pro-cash legislation, because cash jams this integration.

While Amazon still requires me to manually click 'Buy', others are pioneering integrations that enable automatic payments. For example, as you leave an Uber ride, Uber's data centres automatically reach into your bank's datacentres via Visa or Mastercard's data centres and initiate payment. The trend is to consolidate previously separate systems into clusters that can initiate multi-stage and multi-industry processes with one click.

As these styles of integration become ever more advanced and multi-directional, they also become less visible. All major tech companies have announced an official interest in making a foray into finance, but they do this by entering into partnerships with core cloudmoney institutions, so it becomes challenging to see whose systems are being used. For example, Uber announced Uber Cash, which is built on top of major pre-paid card company Green Dot, which plugs into the banking system. Alexa is integrated into Amazon Pay, which is integrated into banks, while Apple has launched Apple Card, backed by Goldman Sachs. Google has announced Google Cashe in partnership with Citigroup, while Facebook continues with its payment projects (a case we shall return to later). J. P. Morgan stands behind AirBnB and Amazon to provide new connections into the banking system, while in India Paytm integrates payment with e-commerce, and in China WeChat and Alibaba do the same.

But if the initial drive is to create corporate clusters, the next is to tether people to them in ever more advanced ways, such as via biometrics. Alipay and WeChat in China have pioneered payment by facial recognition – staring at the camera triggers a money movement between institutions. All cloudmoney institutions are excited about this drive to fuse your body into their overall structure. This is something long predicted in science fiction: in Marge Piercy's *Body of Glass* (1991), payments are triggered by fingerprints (leading to a class of criminals who steal access to bank accounts by cutting off people's hands).

Even objects can be tethered to the cluster by IoT – or Internet of Things – infrastructure. Firms are excited by the idea of connecting objects to accounts, so that those objects can act as agents standing between people and companies. This principle is at play when tags placed on cars trigger payments as they pass toll roads, or when Amazon Alexa initiates payments between your bank and Amazon. All manner of hybrids can be dreamed up, like fridges that can buy milk from passing drones.

What happened to the bankers?

At a ground level, the fintech sector might express itself as a flourishing of apps, chatbots, devices, wearables and biometrics, along with Big Tech integrations, but these all lead to the same core 'operating system'. They dock into the underlying banking sector, rather than bypassing it, and frequently they dock in at multiple points: new-wave apps that offer suites of banking services, FX services, and wealth management services do so by plugging in to multiple background financial institutions. Thus, the model in which I speak with a branch manager who decides upon a loan is replaced with me filling in data on a smartphone app that plugs – somewhere down the line – into the back-end IT systems of various members of the banking oligopoly.

This leads to a new question. We are seeing the rise of self-service interfaces that mimic personality, but where do those lead you to? In industry jargon the different interfaces are called 'channels' – they channel you in, each setting in motion a 'user journey'. The fifteenth-century user journey was a walk down the street to a building where the assistant would lead you inside to a stern-faced lender who told you his terms. The twenty-first-century 'journey', however, is beginning to feel like entering a hall of mirrors that rearrange themselves depending on who knocks. In this world, you'll never shake the hand of the financier at the core of the bank because, just as you cannot touch digital money, so you cannot see, or speak to, digital financiers. Let's meet this new generation.

9

Sherlock Holmes and the Strange Case of the Data Ghost

The NASA Ames Research Center in Mountain View, Silicon Valley, is a curious intersection of US military power and techno-utopian capitalism. My sleeping quarters are in a barracks with signed photos of astronauts on the wall, stocked with copies of *Military Spouse* magazine. Across from the parade ground outside is Moon Express Inc. It touts itself as a 'privately funded lunar transportation and data services company establishing new avenues for commercial space activities beyond Earth'. Further on is a huge skeleton of a structure that once housed the USS *Macon* airship, and an old McDonald's now serving as a centre for digitising photos of the 1969 moon landing. I look through its window and see Apollo 11 images above a disused burger grill, while F-16 fighter jets land behind me on Moffett airfield, part of which is leased from NASA by Google.

Here too, in this sprawling compound, is the Singularity University campus, founded by space entrepreneur Peter Diamandis and Ray Kurzweil, Google's prophet of the coming technological 'Singularity'. This is the idea that, through our enmeshed innovations, humans will trigger an 'automation of automation': a tipping point

'McMoon' and the Airship Hangar

at which intelligent machines create other machines and give birth to a vast technological 'super-intelligence' that we can fuse with to become gods of our environment. Such techno-utopians have warned me that Earth is under threat by asteroids, requiring us to accelerate into an interstellar species. The idea that we will become super-beings that transcend the earth echoes Christian thought. The former Christian evangelist Meghan O'Gieblyn argues that Singularity stories mirror biblical salvation stories almost exactly, but replace God with Technology.

The vision creeps me out, but to some in Silicon Valley it provides a utopian tale of technology solving all earthly problems. This is an appealing cover, because venture capitalists mostly back businesses targeting 'first-world problems' (such as 'My taxi took fifteen minutes to arrive' or 'Shopping is inconvenient'). Singularity University was established to focus on how 'exponential' technology built by start-ups could leapfrog over all obstacles to solve more serious problems – including poverty, hunger, disease and . . . death itself (a faculty member informed me the organisation coalesced around immortality research). Wannabe entrepreneurs from around the world are hand-picked to come here, funded by Google, to brainstorm about saving the world through technology.

The vision is a hallmark of the Silicon Valley brand of libertarianism, which sees the striving of entrepreneurs as the source of societal progress. It is an appealing self-image for some, but it runs into opposition from the Marxist tradition. Marxists see entrepreneurial elites as limelight-hogging figureheads, who are nothing without hidden armies of labourers to do the actual work. This was certainly true for nineteenth-century industrialists, whose fear of labour movements led them to promote the discipline of traditional conservatism, ideal for getting workers to fixate on their little piece of the pie without questioning the overall class hierarchy. Twenty-first-century tech firms, however, can afford to appear more countercultural, because they seek to bypass labour armies entirely. They depend instead on much smaller brigades of highly paid professionals – brought into alignment with share options and other perks – to build automated platforms with billions of users.

Despite the self-image of their founders, these platforms have immense capacity to scan and monitor the underclass of society. When I was at Singularity University, Joe Lonsdale of Palantir visited to share entrepreneurial tips with the students. The corporation – named after a crystal ball from *The Lord of the Rings* – provides data surveillance technologies to the US military, CIA, NSA and FBI, and was co-founded by Lonsdale alongside PayPal co-founder Peter Thiel. Palantir emerged from PayPal, because the digital payments giant developed systems to automatically scan through its millions of customers to automate fraud detection, and Thiel then funded Lonsdale and a few others to apply the same model to national security. In addition to providing services to state authorities, however, Palantir rents itself out to major financial corporations like J. P. Morgan.

As he concluded his visit, Lonsdale noted that both he and Thiel were also heavily involved in the Seasteading Institute, a libertarian project to build stateless cities on the sea. That may seem at odds with Palantir's purpose, but Silicon Valley is full of such contradictions. Its market techno-utopianism has always been deeply intertwined with state and military support in the form of subsidies, state-funded R&D, and procurement contracts. And the military-tinged AI technologies being built here are now being co-opted into the service of financial institutions.

Replacing the urinals

The bankers taking advantage of the Commerzbank Tower urinals in Chapter 1 have no way of directly connecting to every single person they see on the street below. Historically, bankers occupying corporate headquarters could only deal with masses indirectly, via the federated front line of branches. In the last chapter, however, we saw how that federated frontline is being replaced with digital interfaces that connect millions of customers *directly* into the core of the banking system.

Most readers will have experienced the various automated options such systems offer, but those systems are collecting enormous amounts of fine-grained data about our interactions in the process. HSBC's interactions with its 39 million customers, for example, have generated some 150 petabytes of data. To picture this, imagine if your 1-terabyte computer was a physical warehouse: your photos are like a thousand dusty paintings stacked in one far corner, while your 600 Word documents are manuscripts packed in a box stored on one solitary shelf. It would take years to fill this warehouse, because it could handle hundreds of thousands of files. A hundred and fifty petabytes is like 153,600 such warehouses, all crammed full to the ceiling.

There is no point gathering data, however, unless it can be made sense of and acted upon, but the employees in the skyscrapers are not going to have the capacity to sift through millions of individual records. That is why companies like Palantir exist. They specialise in automating this process of wading through data and making decisions in response to it, scaling that far beyond the capacities of human analysts and deciders. These companies seek to build predictive machine-learning and AI systems that can make sense of data about people before acting on a particular goal (like 'sell to', 'exclude', 'include' or 'eliminate').

Bank service staff in branches traditionally directed information to managers in central headquarters, but as banks automate away the former they have to simultaneously invest in technologies to augment – or even replace – the latter. They have discovered that more profit can be made by replacing those urinals with server racks, and are racing to build their own in-house AI systems to make decisions. The Royal Bank of Canada, for example, has a team of 100 PhDs researching AI.

Big investment banks were early pioneers in using algorithms to replace core staff. Indeed, in 2019 I watched Goldman Sachs' head of technology at an AI conference in London say, 'We have 11,000 software engineers, 4,000 of which for decades have focused on . . . machine learning . . . When I joined Goldman Sachs twenty years ago we had 200 traders on NASDAQ executing orders. Now we've got three, and it seems quite natural to us that evolution.'

The traders she was referring to are those screaming red-faced men in trading pits, famous for taking in information and making quick decisions about markets in scenes of high macho drama. Their ranks have now been decimated by the introduction of electronic trading platforms (built by those software engineers). This process was chronicled in the 2009 documentary *Floored*, which follows the lives of the legendary pit traders of the Chicago Mercantile Exchange as they are made redundant by a class of people they refer to as the 'computer boys'. These big-talking traders were once patronised by financial mega-corporations and made to feel like heroes. Now they have the air of war veterans thrown out on the street.

Players like Goldman Sachs are behind-the-scenes investment banks that do not have to deal with front line retail customers. They don't have to spin stories about using tech to build nice consumer apps, because they use it in the core of global finance. High street commercial banks, however, have invested much energy into claiming that their automation drives are for the good of the everyday person. But while someone might be OK with inputting requests to a bank via a digital app, very few customers have ever demanded – or expected – that those requests might subsequently be assessed by a non-human banker. Increasingly, though, commercial banks are working on systems to automate everything from loan decisions to insurance claims, and we already know the drill: the sector is going ahead with this regardless of customer expectations.

The evolution of financial robots

To help understand the evolution of this internal process, picture me strumming chords on an acoustic guitar for you by a campfire, using the instrument to turn my body's energy into sound as I strum. Now imagine yourself standing forty metres from a stage where I plug an electric guitar into an amplifier and turn up the volume. This time my strum produces a much louder sound. Rock and roll depends, as it were, on gas and coal, burned to produce electricity.

Now imagine yourself hundreds of kilometres away from me. I plug a synthesiser via an amplifier into a computer and turn it to its 'electric guitar' setting, so that pressing the keys generates guitar sounds without me having to strum. I record a synth-guitar song using software and, with a mouse-click, can hear the entire composition again without having to perform it again. I can then stream it to you and many others via the Internet.

Finally, imagine a hypothetical 'auto-composition' programme that, with a single mouse-click from me, composes and records synth-guitar songs while I remain entirely divorced from the process, and then streams those to you and many others. They might not be very good songs, and may feel somewhat empty, but it is certainly a very efficient system.

This progression from campfire strumming to auto-composing robo-musician illustrates an automation sequence. Notice how the process leads to my input triggering ever-greater output. At one extreme I exert much energy and passion into playing for a single audience member, and at the other a single finger-twitch can create an entire album for millions of strangers to hear simultaneously. The further up the automation chain we move, the more distanced I become from the creative process, and the further that you, the listener, become distanced from me, the supposed 'creator'.

We can use this as a metaphor for the automation occurring within banks. The financial equivalent of 'strumming a chord by the campfire' would be a fifteenth-century Florentine banker writing out loan contracts on linen paper, using an abacus to speed up calculations (thereby providing a kind of 'acoustic amplification', enabling the financier to reach a number of people during the day). The abacus is a manual tool, but to turn it into a machine requires adding non-human power to it. An electronic calculator is one such automated abacus, and this 'electrical amplification' enabled the 1960s British banker to work much faster than our medieval Florentine, and to reach more

159

people. By the 1990s the most ubiquitous financial machine was the spreadsheet, which greatly speeds up laying out and analysing data.

When I worked in high-finance derivatives, we had very elaborate spreadsheets, such as one that would draw on dense demographic data to output the contract terms of a 'longevity derivative' (an obscure bet used by life insurance companies to protect themselves from changes in a population's average life expectancy). This spreadsheet could do in minutes what a person would need days to do manually. Likewise, a credit-scoring model takes in data inputs about people – such as where they live and their income – and outputs a single, quantified score in a matter of seconds. In our metaphor, these models are somewhat like synthesiser keys, which artificially replicate analogue human effort.

Long-distance 'streaming', though, requires communications technology. Traders might use elaborate spreadsheets to amplify their ability to make quick decisions, but it is telecommunications channels that originally enabled them to project those out to others at a distance. Such 'streaming' is constrained by their human limits. There are only so many calls they can make, and so many times they can click a mouse before they need a smoke or toilet break. This is why agency is being transferred to the machines originally used to amplify their efforts, to create *financial robots* with decision-making capacity. An automated trading bot, for example, is akin to an Excel spreadsheet that receives a data feed from the outside world, makes calculations, and then outputs orders back into the world. Similarly, an automated loan system might automatically grant loans to someone who passes a threshold credit score.

Such robo-systems do not need toilet breaks. Furthermore, they can be upgraded into the realm of 'auto-composition' if they are given the capacity to *learn*. This leads to AI: digital systems that learn from past experience to create their own strategies. Like musical auto-composition they may not be particularly good at doing that, but banks see in them great potential for efficiency gains. And, to such a system, the 'past experience' they learn from is just logs of recorded data.

I, Robo-banker

An early portrayal of artificial intelligence is found in Isaac Asimov's sci-fi classic *I, Robot* (1950). In the first chapter we meet Robbie, a simple first-generation robot that cannot speak, but by the book's third chapter we have moved forward in time to meet Cutie, an arrogant speaking robot that can manage a space station. Asimov's robots all have a 'positronic' brain – a complex computer constructed from a mass of circuits – which is inserted into a mechanical body. By the final chapter, however, the most advanced robots no longer have a body: immensely powerful positronic brains (called 'the Machines') sit static like meditating sages, making sense of vast reams of global-scale data impossible for ordinary humans to see patterns within.

This vision of AI – in which powerful 'thinking machines' take in data and attempt to see patterns within it – finds a more down-to-earth parallel in our own experience. For example, as a teenager, I used to birdwatch, and had a notebook to jot down when and where I saw particular species. The birds themselves did not keep logs of their own activity, but in the presence of a watcher like me their behaviour yielded a constant data by-product, and I gradually learned to see patterns in it. We could imagine an AI system that does something similar, constantly scanning the world to learn on a much larger scale, like Asimov's sages.

The key difference, however, is that I collected that data for my own personal curiosity. I did not use it to try influence the birds' feeding patterns, or to alter their migration choices by blocking entry to airspace. Systematised people-watching, on the other hand, usually has an *agenda*. While social scientists may purport to watch people out of pure intellectual curiosity, much of our every-day watching is attached to judgements that inform action, such as, 'That guy looks dodgy, so avoid'. It is the pairing of a *categorisation* ('dodgy') with a *goal* ('keep safe') that yields an *action* ('avoid'). This combo alters depending on who is involved – to a cop, 'dodgy' might yield the action 'approach', because their goal is different. Each element can be questioned: has the right categorisation been made? Has the right course of action been decided upon? And is the goal worthy (this often goes unquestioned)? In the case of bankers the goal is profit, and this in turn informs other questions, such as: 'What kind of person are you?', 'Is it worth giving you an account?', 'Should we lend to you, and on what terms?', 'Are you trying to deceive me?', 'How can I get more business out of you?', or, 'What product is suitable for you?'

In the past, financiers answered these questions by observing you closely, gathering information about you through networks, investigative clues and interviews. They might have even judged your clothes and demeanour as you approached them. You may have been coloured by a past that could be used to predict your future behaviour, or by your association with others. Patterns would be noticed subconsciously, and surface as background intuitions or 'hunches'. Imagine if Sherlock Holmes shifted his goal from crime-fighting to profit-seeking. He could probably size up someone's creditworthiness or profit potential by careful observation of small details like their mannerisms and tones of voice. He could have made an incredible nineteenth-century financier.

When a banker is distanced from the person they are studying, however, and receives detached data, it is harder to produce hunches about who the person is and how they will behave. A modern Sher-lock could still do this, wading through thousands of digital records about the person's payments, for example, which show what they did, when, where and with whom. But Sherlock is an expensive person to hire. It may be worth bringing him in for high-status investigations that involve, for example, tracking down the offshore finance network of a Mafia boss, but not for low-value decisions such as, 'Should we grant account no. 2337569 an overdraft?' To use a related example, YouTube wishes to boost advertising revenues by driving its users to watch more videos, but – relative to how much it makes from each user – it would be far too expensive to hire an actual Sherlock to curate video suggestions. It would also be impractical: there are just not enough Sherlocks in the world to cover all the users. YouTube must build automated Sherlocks, much like banks.

But how do you build such an automated investigator? Well, one way is to get investigators to build a mechanistic model of themselves, based on general experience. If an experienced banker were asked to codify their knowledge, they could attempt to create a formula, such as 'person from this location + this income + this history + credit on these terms = likely to pay back'. In the early days of computers, this could be written out as an algorithm, a recipe given to a computer by an expert that has crafted it. Just like a bread-making machine outputting loaves of bread if given the ingredients, a basic credit-scoring algorithm can output predictions of the likelihood of repayment if fed the right ingredients. Financial institutions can draw upon the data provided by credit bureaus like Experian (which pool data shared by other institutions) to get the ingredients for their proprietary credit-scoring algorithms.

But if such systems began as recipes followed by computers, they are increasingly being taken beyond that by machine learning – the branch of AI that seeks to turn computers into problem-solvers rather than order-followers. There are many different styles of machine learning, much as there are many different styles of human learning. Consider intuitive learning patterns, such as 'Person X displays similar traits to people who previously did Y, therefore may also do Y,' or, 'Last time I touched a hot stove it did X, and I prefer Y, so will not do it again.' In a person these are often intuitive – you do not need to tell a child that they have a pain-avoiding principle, and do not need to explain to them that we see patterns in the world. A computer, however, needs to be told how it will learn. It cannot feel pain and has no ability to assess what success is or why it may wish to discover and act upon something. A computer is a natural order-follower, so you need to order it to become a problem-solver, a process that requires rather advanced coding.

A thought-experiment can convey an approximation of one such style of learning. A traditional bread-making machine is programmed with a fixed recipe, but imagine you were asked to programme a machine that will 'learn' to make bread. The first step would be to programme in a *learning methodology*. But learning always requires *experience*, so the second step would be to feed it data. This machine's learning methodology is thus designed to take in data about what completed bread loaves generally look like, along with data about possible ingredients (e.g. flour, pineapples, asparagus, yeast, water, rice, salt, etc.) and processes (e.g. heating, freezing, mixing etc). Its aim is to find a formula that connects the desired output with the possible inputs, and to then test that and feed back the result as new data. Mixing flour with asparagus and freezing it might not work well, but over time a machine-learning bread-maker trained on hundreds of years of data might be able to infer subtle patterns about bread that humans only have vague intuitions about.

The idea above is fanciful – no such bread-making machine exists – but this is one way to capture the idea of how machine-learning systems are 'trained'. For fintech pioneers who are enthusiastic about this process, it is seen as a strategy for creating financial machines that can extract actionable information out of piles of existing data, and a means to extract from it esoteric *new* sources of data: with enough computing power hundreds of variables might be tested to draw out patterns that a human financier might not see.

A famous feature of the old Sherlock Holmes novels was his ability to make inferences from pieces of information that seemed totally unrelated to the quest at hand. This always involved close observation of individual people. Imagine Sherlock watching a particular Londoner travel from the poorer neighbourhood of Tottenham to upmarket Moorgate in London's financial district. He hypothesises that they might be an office cleaner, but this hypothesis changes when he sees them return to a different stop later that night, and not return to Moorgate in the morning. The new clues suggest the person may have neither permanent abode nor employment. The more he gathers esoteric pieces of information, the more refined his hypothesis about them may become. Large institutions, however, have no interest in trailing individual people like this. Rather, AI systems are set to work like statistical sniffer dogs to uncover correlations within *mass* data, to generate hypotheses like 'People who travel from this area to this area at this time have low default risk.' That can be tested, and the results fed back over time to refine it.

An employee at a fraud-detection fintech start-up told me his company uses over 500 data sources from a phone to assess you. This could include, for example, the speed at which a person types. Indeed, our phones send all manner of intel on us back to institutional datacentres, which can be pooled with the data of others and mined for clues to generate hypotheses. This returns us to the theme of the previous chapter, in which apps hosted on phones 'stick' to us like a spy. The app might query your device to see what other apps you have, interrogate your contact list for a register of your associates, and pore over maps of your walking patterns. And even when banks do not have such a direct portal into your world, they can source data from elsewhere: after all, our every website click and contactless payment card tap is logged. These are the sources of information building the new generation of robo-banker systems.

No shit, Sherlock?

The real Sherlock loves two things. First, solving a case, and secondly, being able to recount it to you as a story. Imagine, for example, he discovers a person in Kenya with two Facebook profiles. Having studied the case, he tells you that they are a gay person presenting a straight face to their conservative family with one profile, and another to their sympathetic friends. 'Robo-Sherlocks', on the other hand, are not so good at storytelling. They have no ability to mull over the nuances of a single person's case, or give you a blow-by-blow narrative of how they came to an assessment. Perhaps the system has uncovered a correlation between multiple Facebook profiles and fraudsters who set up fake profiles, so once Robo-Sherlock 'solves' the case, they simply say 'Person is fraudster', rather than 'Person has devised a double-life to conceal their true self from their intolerant family.'

While old algorithmic systems have a known recipe, you cannot ask a multi-dimensional AI system 'why' it thinks something. Imagine asking Spotify or YouTube why they suggested a particular song or video to you. Their systems cannot 'hear' or 'see' the songs or videos. Rather, they collect data from millions of users and cross-reference it to create statistical clusters of songs and people that tend to overlap. In the realm of music suggestion that may produce some beautiful synergies (or maybe get you trapped in a bubble),

166

but in the realm of finance it may really matter if you find yourself in a Kafkaesque doom-loop of mis-categorisation.

And there are countless possible mis-categorisations. To use a light-hearted example, I discovered that Twitter's Robo-Sherlocks have me categorised – among other things – as a possible 'working-class mom'. In Twitter's case this may mean I'm shown adverts for cut-price family holidays, but in the case of a bank it could mean getting blacklisted from credit, or even its opposite – getting enticed into over-indebtedness. Regardless, if bluntly auto-categorising can help a bank make money by allowing it to churn through many more customers at low cost, the bank will try to do just that. Much in the way YouTube does not care if it gets video suggestions wrong for 20 per cent of people, so too will banks write off those customers who find themselves punished by the imprecision of their systems. The motivation in automating 'intelligence' is not to seek the truth and nuance of every case: it is to optimise profit at scale, and, unless it hurts the bottom line, financial institutions will not discard a system if they discover it only has the capabilities of a lazy holiday intern.

But imagine if this automated 'lazy intern' had bigoted parents, which in the case of AI means models that encode the hidden biases of the creators, or models that get fed training data forged in the context of structural racism. Loan denial rates in the US are historically far higher for black people, and if such a skewed data set can be allowed to calibrate the mind of the AI 'child', there is a risk of automating the biases of the status quo. On the other hand, financial institutions increasingly claim that AI systems can be used for financial inclusion, helping them to lower the costs and risks of dealing with more marginalised groups. In Chapters 6–8 I argued that financial inclusion can be better thought of as absorption; and as digital payments spread into the peripheries, they are followed by the AI technologies.

Inclusion through surveillance

In 2016 I watched a man from a major financial inclusion institution speak at an IMF-sponsored event in Singapore. With excitement he explained that his organisation was experimenting with the idea of spying on poorer Ugandans' mobile phone travel data to work out their creditworthiness. I asked whether this surveillance system could also be hijacked by regimes looking to punish political opponents. After all, Privacy International reported in 2015 that the Ugandan government was pioneering all manner of surveillance techniques to monitor political dissidents. The speaker mumbled a mantra about people needing credit. If financial institutions must spy to be profitable, then so be it.

Since the 1980s it has become increasingly fashionable to assert that people can be raised out of poverty by giving them access to credit. This has resulted in the field of microfinance, which rests on large numbers of poor customers being given tiny loans. Its foundations have always been shaky, because the world's poorest people often occupy the lowest rungs in commodity-producing countries run by feeble governments that are subservient to more powerful advanced industrial countries. In this context, a person's lack of income stems from being in a weak position within a transnational economy, and there is no guarantee that giving them credit will magic away the reality of that situation. Credit is empowering when a person is in a strong position to generate future returns that can be used to repay the loan, but it is not empowering in and of itself.

There are clearly positive cases of microfinance, but the overall paradigm has often led to over-indebtedness. But rather than questioning the paradigm, techno-optimists seek to use digital technology and AI to refine credit-scoring while expanding credit. People are denied access to credit for one of three reasons: their numbers do not add up; there are no numbers on them; or there is no record of them to attach numbers to. The first step in any credit expansion project is to deal with the third reason, which is why the Indian government, for example, established the Aadhar identity programme. The programme collects people's biometric details such as fingerprints and iris scans and uploads them to a central database that can be used by a range of institutions to verify a person's identity. This in turn allows different parties to start building a dossier about a previously undocumented person.

A host of start-ups now focus on how to build unorthodox intel on these so-called 'thin-file' or 'no-file' clients. For example, banks buy the services of companies like Lenddo, which analyse the 'digital footprint' left by someone's phone usage. They can sift through a person's social media activity, browsing activity, location data, contacts, call history, installed apps, calendar events and phone model. The company claims to use machine-learning to analyse up to 12,000 variables to work out a credit score in less than three minutes, which is then fed in to augment a bank's own credit scoring systems. Companies like Tala in Kenya analyse 10,000 data points from a wannabe borrower's phone, including how often they call their mother, and whether or not they prefer car-racing or zombie-themed mobile games. Newer start-ups like Neener Analytics apply psychometric testing techniques to 'undecisionable' people (industry jargon for people that banks struggle to make a decision about). Built by two men, Neener's 'automated psychologist' system manifests in the form of a female chatbot called Aria, which asks subtle questions to psychologically profile people, and assess them 'not based on what the person talks about, but on how they talk about those things'.

These Robo-Sherlock systems are attractive to financial institutions in the low-income realm because the profit margins for each customer are much thinner. As a result financial inclusion has become strongly associated with digital automation. Poorer people are unprofitable if an institution has to spend too much time serving or thinking about them, so the solution is to automate the process of doing both.

The return of Big Bouncer and Butler

In Asimov's *I, Robot*, the AIs are built in such a way that they cannot act upon orders to harm humans, or to give results that harm humans. They have noble intent programmed into them, and are required to apply it to humankind in general, rather than using it to benefit specific interest groups. Financial AI systems, however, not only carry the potential to be incompetent, biased and unaccountable, but are also employed to serve institutions, rather than humankind in general. In Chapter 6 I introduced the concept of Big Bouncers and Big Butlers, and AI is used for both. Banks employ AI in Big Bouncer roles to decide upon access, and in the context of any cashless society – where digital participation is forced – finding yourself blacklisted on shared digital credit-scoring or fraud detection systems could mean being shut out of the economy.

But Big Butler looms large too. If I spend an hour watching the surfing videos that YouTube suggests I might like, the system does not interrupt me with a pop-up that says 'You seem to enjoy surfing, so why not go outside and swim in the ocean?' The goal of the system is to sell advertising, which means YouTube's 'helpfulness' is designed to keep me locked in with more videos (like helpfully offering alcohol to an alcoholic upon observing their past behaviour of drinking alcohol). Likewise, financial institutions do not spend billions investing in systems that analyse our payments data as an altruistic act to help us achieve our holistic life goals. Rather, analysing behaviour is used to drive product selling and keep people locked within their ecosystems. This is the Big Butler function of AI systems.

Big Butler systems are disturbing because they can create strange feedback loops. Imagine if I, as a birdwatcher, fed my notebook data through models that in turn informed the placement of enclosed aviaries. Imagine the data is used to position movable bird feeders within those aviaries, and to trigger speakers that mimic the birds' calls to induce wild birds to fly in, after which I could continue to watch them. All my notebook entries captured after this would be polluted by the fact that I had rearranged the environment. I am watching them under captivity.

In the digital world these enclosed aviaries are becoming real, albeit hard to see. You might think you're browsing a public website, when in fact it's private: everything you see on it is tinted by your previous actions, and the tints get darker as they feed back on themselves in self-reinforcing loops. The institutions that control these views can dim certain elements and highlight others, while auctioning off rights to third-party institutions who wish to curate the view.

The outward re-skinning of finance and its integration with AI systems raises the spectre of similarly artificial environments, within which micro-monitored customers can be induced into changing their financial behaviour, spending patterns and borrowing. For example, insurance companies are excited about *dynamic pricing*, in which prices cease to be standardised and become tailored to whoever views the good or service in question, eroding the old insurance principle of *mutuality*, in which people gathered into collective groups to offset the risk each one faced alone. In old market economics the 'market price' is supposed to be a publicly visible reflection of the average will of buyers and sellers – everyone sees the same price – but if we are steered into privately curated bubbles by institutions that have intimate data about us, that begins to break down.

We are increasingly haunted by data collected on us by automated people-watchers, who use it to alter the landscape we see, the price we are presented with, and even the sound of an automated debt-collection agent's voice we hear (when that automated loan goes wrong).

Placed in the hands of a philosopher, these bodies of data might be used to reflect on the multi-faceted nature of humankind. But when held by institutions with their own agendas, data accumulations become 'possessed' with a singular intent, and hover over us like data ghosts blocking the way or pushing upon us. And while they may have legitimate uses, as these data ghosts proliferate we sense their presence and perhaps become nervous of them. In situations where access is based on an invasive swathe of dynamic micro-data, panopticonism can creep into your life. What new data will emerge from what you buy, or who you hang out with, or how fast you walk? You may soon fret about whether your spontaneous decision to take the day off to sit on a beach will manifest as a data-point in a mystifying mesh of relations that ultimately leads to a declining credit score.

10

Clash of the Leviathans

The old vision of capitalism as an economic system within which ordinary people bargain on markets is beginning to feel quaint and outdated. It can even feel romantic. Think of classics like Jack Kerouac's 1957 novel *On the Road*. It is set within a mid-twentieth-century industrial economy, where lonely individuals work precarious jobs in big cities. They hustle for cash in the market economy, spend it on beat-up cars and speed down America's highways. On the road, they cross paths with other yearning souls for brief moments of intimacy, before moving on again.

The world of this novel has remarkably little bearing on today's reality. In the society it portrays you can hand over cash at a second-hand car dealer's for a rusty Pontiac whose only tethering to the distant world is a radio receiver. Big institutions exist, but they do not infringe on every aspect of life. The characters know how to fix a car, and don't care that their actions aren't garnering likes on social media. Their adventures are recorded in personal memory only, and do not accrue upon a database to form a data ghost that will be used to personalise their environment, or influence their price of credit.

When we take stock of our current situation, it would appear that the capitalism of the future does not want to let you pay with cash or drive off on your own. Instead, corporations are encroaching on your private space as the car reports your movements to the Cloud. The characters in *On the Road 2030* have cars that automatically pay for toll-roads, correct them when they take the wrong route, and are tied into an omnipresent market for data: as the pedal hits the metal the car either automatically slows down, or automatically increases the driver's insurance payments. There are no unpredictable paths, and no bumping into randoms on the road of life.

Perhaps our future driver has an automated bank loan secured against their car, policed by a remote kill switch that renders the car unmovable if their balance gets too low. Regardless of the details, the overwhelming drive in modern capitalism is to build systems that run on autopilot, the people within them increasingly behaving like passive observers.

Innovations that facilitate such a future are pitched to venture capitalists every day, and every day they get funded. The role of those investors is to finance the development of small parts of an overall control complex. For ten years I have received daily mailing-list updates about breaking fintech news, seeing stories of thousands of start-up companies passing through my inbox. On a daily basis the evolution of the stories seems imperceptible, but over a span of years becomes obvious: what begins as a story about a start-up raising early seed-funding turns into a story of the same company's acquisition by a bank, or its new partnership with Visa or Amazon. The ones that do not manage to integrate into corporate capitalism like this disappear, turning into archived emails that slowly fade from consciousness.

There may be occasional rearranging of relative power between the players, and the rise of some new names – like Square or Stripe – but when we cut through the product pitches and catchy start-up names, the pattern remains: the financial giants are being outwardly re-skinned, while their inner core is replaced with profit-seeking machine intelligence, and, as this process advances, it unlocks ever-more 'sci-fi' imaginations. Entrepreneurs must constantly push at the frontiers of what is considered acceptable, because they are channelling the will of investors desperate to find the next big thing within an economy that must expand. They will throw the driver out of the car too, if it promises to be profitable.

But while these trends I describe are real, they have not expressed themselves equally around the world. I have worked in rural parts of South Africa that feel like the Wild West – the electrical grid constantly breaks down, there is no phone reception, the state feels distant and tribal elders and kinship groups have more power than local magistrates. In these places there are no venture capitalists or chain stores with contactless card terminals. Rather, battered physical cash gets passed around in outpost shops, while subsistence farming and pastoral cattle herding continues. In fact, in rural South Africa you might even imagine Jack Kerouac's 1950s world as you fix your overheated car on a patch of crumbling highway, with no mobile Internet to help you.

The slow absorption of rural South Africa into transnational surveillance capitalism is a process that will take some time, but will lead to the end of the Kerouac world there too. This ongoing process is the latest iteration of a repetitive cycle in which capitalist systems shed previous versions of themselves in order to expand. We saw an earlier iteration of this in colonial times, when European powers attempted to absorb entire regions and force them into the global economy. This created uneven zones of overlap, the legacy of which continues to this day. In South Africa, for example, the meeting between monotheistic European Christianity and pantheistic Xhosa spirit worship produced hybrid syncretic religions that blend the two together; similarly, the expansion of the global economy has also left a legacy of 'economic syncretism', where older, informal and more localised economies mix with transnational ones. This is what a tourist from Denmark experiences when walking through the noisy street markets of Amazonian Pucallpa, where knock-offs of international brands will be found in unregulated stalls alongside people selling herbal remedies and alligator steaks cooked on open fires. The quintessential image of a 'developing country' is one of syncretic mixes between the traditional and the modern. They are places where formal institutions are thinly spread, such that you have to deal directly with people, or bargain for stuff rather than being dictated the price by a supermarket executive. The implication is that full 'development' is reached when the syncretism disappears, and the street markets are fully replaced with malls selling transnational brands.

This hints at something important. So far this book has focused on the expansionary actions of big companies, but capitalist markets are built upon a substrate of states tied within complex webs of geopolitics that date back to colonial times, and before. Let's take a step back to tease out some of the interactions between states and markets at a global scale, as these powers form the backdrop against which a new transnational rebellion – spearheaded by cryptocurrency – is forming.

Leviathans, and their conflicted alliances

The relationship between states and markets – locally and internationally – is complex and contradictory, so let us pick our way carefully through some key points. The first point is that *informal markets* can find ways to operate syncretically in the shadows of their local states while drawing upon state infrastructure, such as the cash system. This is the image I described above, of noisy vendors selling unregulated goods in street markets.

The second point is that formal state institutions begin to permeate back through informal markets to create more modern, semi-informal markets: for example, Brixton Market in London was historically a place where people from ethnic minority backgrounds carved out an under-the-radar existence in the UK, but the Jamaican merchant here now has an alcohol licence and a hygiene standards certificate pinned to the wall. This is because this market is increasingly subject to centralised regulations. This in turn can help decentralised market activities proliferate, because those regulations neutralise or dampen those aspects of our relationships that could stand in the way of commerce, such as lack of trust or lack of common standards. If the merchant sells me dodgy products I can sue him using the state legal system, but if I run out without paying, he can call down the wrath of the police on me.

As a result, strangers can do business with each other without either party fearing the other, which enables the markets to *scale* to much larger sizes. That formal state institutions catalyse *bigger markets* was hinted at by Thomas Hobbes in his *Leviathan* (1651). A state 'leviathan' is a set of institutions that will strike you down if you breach property rights or break contracts, and serves as a central body via which large groups of people resolve disputes. Such an entity unlocks corporate capitalism, because larger-scale corporate markets depend on extensive and strong systems of company law, contract law and private property.

The third point, then, is that state power gives rise to corporate power. Indeed, powerful state leviathans underpin a broader *leviathan complex* that extends beyond the state: modern societies have us subject to common public-private monetary systems run by banks, alongside telecommunications companies, utility companies and – increasingly – mega-corporations like Amazon and Uber that all come to act like second-tier leviathans. These institutions explicitly position themselves as arbitrators that stand between you and me to facilitate and regulate interactions. The large-scale players cannot be seen separately from states and are best understood as different parts of a governing coalition. A major element of modern politics concerns how much the state part of the leviathan complex should assist, challenge or control the corporate part. In the Chinese variant, for example, the state has greater oversight of the corporates, whereas the American variant gives them more autonomy.

The fourth point, then, is that corporate expansion and international geopolitics cannot be seen separately. This is long established: colonialism, for example, revolved around state-backed corporations like the British East India Company seeking new markets, and using their power to extract advantages in foreign regions. Similarly, all the innovations described in this book – from digital payments to fintech and AI – are pioneered and pushed forward by powerful financial and technology firms rooted in strong states. American, European and Chinese commercial giants are supported in their external expansion by their governments, which assist them in reaching out into transnational supply chains and globalised markets. This observation runs in direct contrast to the view that states are pitted against 'the market'. The real tension is not between a generic state and a generic market: it is between, for example, the US state and the Kenyan informal market. In the eyes of the former, that the latter may be resistant to using Mastercard means it is at risk of being captured by a Chinese payment company instead. As we've seen, the US development agency USAID is concerned both with digitising economies and with partnering with Visa and Google to do so. It is not interested in working with Alibaba, Tencent, Huawei or Baidu.

On to the final point. In Chapter 1 I sketched out how Wall Street capitalises big companies who in turn send money rippling through the international payments system like neural impulses. When those international companies are trying to reach out into global supply chains and markets, they come up against a zone of slower conductivity in those pockets of informality where cash is used. Thus, in order to do business in Nairobi, Uber has to allow its Kenyan drivers to accept cash, but the Kenyan cash system jars with Uber's planetary ambitions. If they could flick a switch to replace it with US-controlled Mastercard transfers, they would. They want to see payments crackling automatically from the streets of Nairobi, via the digital payments leviathans, to the streets of San Francisco. And so the US state, for example, ends up pitted against the Kenyan informal market, via USAID.

The same can be said of the myriad commercial partnerships being established by the Chinese government, among others. They all have an interest in dissolving informal peer-to-peer relations, and sometimes attempt to do this by outward-facing state legislation. For example, *hawala* systems run by local Islamic merchants once offered an informal international payments system run via word-of-mouth trust networks: a member of the Somali diaspora in London might hand cash to a *hawaladar* agent in Streatham, who would call up their associate in Mombasa, and tell them to hand shillings to a recipient there. These arrangements, traditionally regulated by honour codes rather than legal systems, are being pressurised by legislation like the US PATRIOT Act, which attempts to illegalise them in favour of bank digital transfers, or, alternatively, to bring them into the formal banking sector.

But even as big corporates use their states to extend outwards, they can lose solidarity with them too. We thus see a complex array of tensions playing out between older leviathans and newer 'techno-leviathans'. Max Weber used the term 'the iron cage of bureaucracy' to refer to the former, alluding to the old government buildings with their drab offices, and the industrial companies with their mountains of filing cabinets. The twenty-first-century levia-thans, though, want to transform the bars of the iron cage into a fine transnational digital mesh. This is what we are currently expe-riencing in the creep of automated surveillance capitalism, which is given different names depending on where it creeps: in cities it is called 'smart cities', in our homes 'smart homes', in our bodies 'self-tracking', and in developing countries, 'digital inclusion'.

A transnational mesh of contradiction

How should one feel about the complex web of relations underpin-ning this growing digital mesh that surrounds us? I have already suggested that for many people it may be psychologically easier to go with the flow and see the rising techno-leviathans as a source of great progress. Certainly, there is a whole range of beautiful things that emerges from interconnected digital technology, from the ability to speak to loved ones over video chat to the ability to capture a favour-ite song through the Shazam app. Internet firms encourage this utopian imagination by continuing to draw upon the optimistic visions of a connected world that accompanied the Internet of the 1990s. Back then the Internet generated visions of a future meta-state, meta-society, meta-market called *cyberspace*. This set the scene for optimistic ideologies like the so-called 'California Ideology', the free-market techno-utopianism associated with Silicon Valley. This is a spirit I referred to in the last chapter, which comes bundled with visions of the Singularity and transhumanism – the belief that humans will transcend their limits by fusing into godlike union with global technology networks.

The modern Internet is not the 1990s Internet, and there are now fears about how it may be facilitating centralisation of power and surveillance. People are noticing the sense of disorientation brought on by information overload, as well as anxiety and polarisation. Awareness of this has risen in the wake of Covid-19, which forced us further into the transnational digital realm. As Amazon packages pile up, and stories of dystopian state facial recognition systems proliferate, many feel a sense of emptiness and concern when thinking of this seemingly inescapable digital mesh.

However, given that people fear the failure of these new technologies they are dependent upon, they struggle to think critically about them. Much like the 'need' for cigarettes only occurs after it has recalibrated someone's body chemistry into a state of addiction, the imagined 'need' for many new technologies often only occurs *after* they have recalibrated our environment. As a former smoker, I can verify that an addict does not light up to experience joy. They light up to avoid the pain of withdrawal symptoms. Similarly, it is largely a dead-end to try and analyse technology based on a notion that people gain ever more joy as it advances. Companies rely upon that narrative to market products, but it's safer to ignore that spin, and instead analyse the constant resetting of expectations we have with each technological change, and the trade-offs built into it. Primary among these trade-offs in modern digital surveillance capitalism is that the technologies we are increasingly addicted to are permanently and directly plugged into powerful corporations and governments.

As the process of tying people into platforms and digital money systems spreads, a person will struggle if they don't agree to be tethered to the market's core institutions. They may enter the ranks of 'the unbanked' and be spoken of as in need of rescuing, and – given their ability to interact with society may be compromised by their lack of tying-in – they may come to feel grateful when they are finally included by the banking sector. But it is those mega-institutions that have forged the connections that have led to our dependency on a transnational digital economy to begin with, and what makes them useful at this scale is also what makes them powerful and dangerous. At scale it is more efficient to have five large banks dominate a national payments system than it is to have 5,000 small ones. Similarly, those national banking clusters find it more efficient to route through one or two international hubs – like the US dollar system, or yuan system – rather than having to build hundreds of direct relationships with each other. The greater the scale, the greater the size of the institutions that can flourish, and the larger those become relative to individuals.

As each bank becomes a touchpoint for tens of millions of people to enter the transnational system, both their power and responsibility increase. A bank may have 30 million different account holders, controlling 60 million devices, who are collectively sending millions of messages each day to try to change their central servers. That is a colossal number of requests, which is why they are turning to those militaristic AI technologies to scan people. They require a robust infrastructure to ensure that the messages sent are not overheard, corrupted, misinterpreted, fraudulent or compromised. They must check whether the requester is the true account holder, which comes with a requirement to identify people and build up profiles of their character, which gives them access to our private life.

Consider the following dilemma. I am a chancer with £1,000 recorded in my account. I devise a cunning scheme to deceive my bank. I open two e-commerce apps on two smartphones, and line up two products that both cost £1,000. I then simultaneously click 'buy' on both products, in the hope that both signals reach the bank's datacentre at the same time to trick it into authorising both. To defeat such an attempt at 'double-spending', the bank is supposed to check which request arrives first and to accept only that one, while rejecting the other. Even if I temporarily tricked them, they control their private database, and can retroactively 'undo' a payment and tell the second vendor to freeze shipping. The entire structure is based on centralised systems of overt access control, with banks acting as chaperones. That may feel good if your bank uses that power to reverse a fraudulent transaction that negatively impacted you, but it is these same features that create all the concerns about surveillance, censorship, exclusion and centralisation of power.

Is there a way to resolve such a set of contradictions and regulate that centralisation of power? The traditional path is for campaigners to lobby government institutions to regulate financial and corporate institutions, but what if you do not believe in those institutions? This is where our narrative takes a turn into the world of cypherpunks, whose technologies are currently introducing fascinating new possibilities into how we arrange our monetary systems.

The cyber-resistance

In the early days of the Internet, the fear of looming conglomerations of corporate and state leviathans led to the emergence of so-called cypherpunks and crypto-anarchists. Starting out in the early 1990s, they foresaw the spread of digital surveillance and began an urgent quest to use *cryptography* – the military art of sending and verifying secret messages – to create autonomous Internet communities. Cypherpunks went on to spearhead divergent movements (WikiLeaks' Julian Assange was a member) but they also set out to pioneer anonymous digital money. David Chaum (whom we met on page 106), for example, saw dystopian potentials in a future cashless society and proposed a system called DigiCash as a private layer grafted over the normal bank system.

The cypherpunks drew upon a range of political traditions, and were an offshoot of radical hacker culture. This is a loose term to describe people with a rebellious attitude towards the technologies and systems associated with large-scale bureaucracy. Hackers like to get behind the user interfaces of these systems, and to explore and subvert the hidden codes within. One popular representation of this culture in recent years has been the series *Mr Robot*, in which a young hacker aims to bring down an aggressive corporation he calls Evil-Corp. The hackers in the series are both anti-corporate and anti-state, an orientation historically associated with left-wing anarchism. Many modern hacker collectives carry elements of that ideology, seeking an alternative non-corporate Internet. There are, however, also strains of right-leaning hacker culture. The latter combines anti-state sentiment with market techno-utopianism to come up with 'anarcho-capitalist' visions of digital free markets.

Anarcho-capitalism is an extreme version of conservative libertarianism, which itself is like the hot-headed younger brother of mainstream political conservatism: louder, looser and calling for more extreme anti-state stances. Libertarians want free markets with a bare minimum of state power to protect property, while hardcore anarcho-capitalists believe that large-scale capitalist markets can survive independently of state institutions like police and law courts. These philosophies could find common cause with cypherpunk movements, which were drawn together by the prospect of creating a 'land of the free' in cyberspace.

The cypherpunk movement was thus politically mixed, but technologically pioneering. Projects like DigiCash ultimately failed, but the various component technologies pioneered by people like Chaum were to become pieces of a puzzle waiting to be brought together into something altogether more powerful, if only someone could arrange them. That someone emerged in 2008. Operating under the pseudonym of Satoshi Nakamoto, they posted a PDF document onto a cypherpunk mailing list as the world's banking leviathans convulsed during the global financial crisis.

The document was short and elegant. Like a puzzle that takes ages to complete but looks so simple once finished, it combined decades of technological innovations within one recipe. It was titled 'Bitcoin: A Peer-to-Peer Electronic Cash System', and has since become known simply as the Bitcoin Whitepaper. It was to become the founding document of the blockchain technology movement, which came with a single agenda: *Replace the leviathans, starting with the monetary leviathans.* In its place would come a new crypto-leviathan, controlled by nobody but usable by everybody. It sounded utopian to some, but as always contradictions abound. It is to that troubled 'revolution' that we now turn.

11

A Paranormalist's Guide to the Spectre of Bitcoin

The Bitcoin White Paper is a nine-page document released onto the Internet in 2008 and authored by an unknown individual called Satoshi Nakamoto. It sketches out an elegant blueprint for how a global network of strangers can collectively co-ordinate and bring to life a system that will issue movable digital tokens. At first glance that might look like a plan for a monetary system, but it is a *token system*, and 'token' is a far more generic concept than 'money token'. The best starting position for understanding Bitcoin is to see it as a token system that some people are attempting to turn into a monetary system. Before approaching it, however, we need to prime ourselves, because it is easy to get misled by the mythology, branding and hype that surround it. To do this we must take a quick detour into the esoteric realm of numbers.

Numbers have the strange property of being able to serve as both adjectives and nouns. In everyday usage most people use the adjective form – 'I will be there in fifteen minutes', 'This is six kilograms', 'I'd like two pies please', 'I have given you three warnings already'. In all these cases the number points to *something beyond itself* (minutes, kilograms, pies, warnings). Mathematicians, by contrast, are unique in that they use 'numerical nouns'. Consider a sentence like 'Fifteen is greater than six, which divided by two gives us three.' In this case the numbers are self-contained objects. They are like the lead actors, rather than a supporting cast.

So which of these two categories of number exist within our monetary system? In Chapter 3 I sketched out the parable of the Giant in the Mountain. In it, the giant gives people a reason to seek out his tickets, after which those become the basis for a broader system of exchange. Those tickets are movable tokens, but they are a specific sub-class of token called an IOU. An IOU might take the simple form of a circular object with the number '1' written on it, but that 1 denotes that it guarantees access to one dose of magical spring water far away up in the mountain. It is not 'just a number'. Similarly, modern money might appear to us as movable objects with numbers, or as numbers on a screen, but those numbers are – in the first instance – *adjectives* denoting a quantity of state, bank or corporate-issued IOUs.

It can be a struggle to see this, and it's easy to mistake the numerical adjectives in a monetary system for numerical *nouns*. We would never say three dogs are 'just numbers', but many people *will* say that about a quantity of money, as if its numbers were mere fictions, rather than legally enforceable accounting records.

The surface-level similarity between numerical adjectives and nouns means we can easily conjure 'money-like' imagery by writing out numbers. If you go into a primary school and instruct the kids to invent a money system from scratch, they will cut pieces of paper into rectangles or circles and then decorate those with numbers. Adults do it too. I can engrave '1' on a flattened bottle top and then jokingly say 'I made a currency', but the number printed on this 'coin' does not refer to anything beyond itself. I could try to claim that it refers to the body it is printed on – there is *one* bottle top – or perhaps the object's weight, but in reality I have just written out a number.

So what does this have to do with Bitcoin? The extraordinary story of Bitcoin is one of an entire movement, now numbering tens of millions of people, that has built a highly sophisticated system for arduously carving out numbered digital objects – the digital equivalent of me meticulously engraving '1' on a bottle-top – and now faces the question of how to turn those into a true monetary system. The stakes are high, because for over a decade now it has been claimed that this offers us the key to escaping the surveillance and control of the banking giants.

Bitcoin paranormalism

To enter an early Bitcoin meet-up – say around 2010 – felt like being a paranormalist detective in a room full of excited witnesses who claimed to have seen a ghost, but all had slightly different accounts of its appearance. Imagine this detective taking down notes, trying to pick out recurring words from the crowd. The four that stand out most prominently are *distributed, decentralised, digital, database.*

Sometimes 'distributed' morphs into 'shared', while 'decentralised' is paired with 'tamper-proof', 'censorship-resistant', 'immutable' and 'incorruptible'. 'Cryptographic' is prominent, while 'database' blends with 'ledger' or 'data structure'. Witnesses shout these out in conglomerations, like 'distributed cryptographically-secured database!', 'immutable shared ledger!', or 'decentralised censorship-resistant cryptographic data structure!' Other nebulous descriptions like 'permissionless' and 'trustless' are thrown in alongside jargon words like 'hashing', 'proof-of-work', 'Merkle tree', 'miners', 'public-private key cryptography', 'digital signatures' and 'consensus algorithm'.

All these words lead to one meta word, which is 'blockchain'. Bitcoin is a token system built using a blockchain, which – according to the detective's notes – is a distributed, shared, decentralised, tamper-proof, censorship-resistant, immutable, incorruptible, cryptographic, trustless, permissionless database ledger, constructed through the intersection of all those other jargon words.

While the technically minded witnesses struggle to describe Bitcoin's blockchain in anything but this incomprehensible mumbo-jumbo, others are attracted to shouting catchy political slogans about how the ghost stands for freedom and innovation. They also are intent upon telling the detective what it *does not* look like. They contrast it with the centralised database, the banks, corruption, flimsy politics, filthy records stored in filing cabinets, governments, Mastercard, third parties, crony capitalism, socialist dictators, inflation, taxation, theft!

One slogan really stands out: *Bitcoin is digital gold, backed by mathematics*. It sounds impressive, but on second thought the detective – me – realises it is not particularly informative. Mathematics is an abstract language, and while I might construct equations to model gravity, that ancient force is not *backed* by a human-invented number system. The phrase seems more poetic than literal: the witnesses are saying Bitcoin obeys rules – embedded in a network of computers – that have a low probability of being broken.

My account is tongue-in-cheek, but this is how many people experience their first encounter with Bitcoin. They are confronted with a morass of technical jargon and political claims, which pertain to tokens that are supposedly 'mined' out of cyberspace and which are also 'money'. It is a starting point that can leave you more confused than enlightened. But if there is one thing I have learned from being a crypto-paranormalist for more than ten years, it is this: glimpsing the spectre of blockchain technology is like building a Rubik's cube in your mind. To get the full picture requires twisting its components around in your head many times in confusion, until, finally, you gain enough experience to see how they align.

A two-pronged attack

Before starting those twists, though, we need an intuition of where we are trying to get to. In previous chapters we considered two key features of the cloudmoney system: first, it is run by oligopolies of banks, which – secondly – preside over a *changeable* global money supply. Bitcoin is best understood as an attempt to provide an alternative to both of those features, by attacking both the centralisation of that system, and its changeability.

Bitcoin promised to provide a means for people to move numbered tokens among themselves without the involvement of the banking sector. This, it was imagined, would pave the way for a *decentralised* monetary system to emerge (under the assumption that numbered objects could indeed be turned into a monetary system, but we will get to that). These numbered tokens, however, were going to have a *fixed supply*. Satoshi Nakamoto – the pseudonymous founder of the system – did not seem to like the fact that banks and governments use their centralised systems to unpredictably issue fluctuating amounts of chips, so desired a decentralised system with an unchangeable supply of tokens, slowly released in a predictable fashion.

These features are in tension with each other. *Decentralisation* is a demand historically associated with a wide range of political groups, from left-wing anarchists and hippy communes to libertarian ranchers and the tribespeople of Papua New Guinea. In the aftermath of the 2008 financial crisis, it was this decentralisation aspect that appealed to both left-leaning activists – who saw banks as domineering private corporations – and right-leaning people, who saw banks as state cronies. Both could appreciate the promise of a system designed to bypass banks, but the promise of *fixed money supplies* is far more politically controversial, because the demand for monetary constraint is traditionally associated with a certain brand of conservative politics. Historically it is creditors – people with a surplus of money – who are inclined to see tight money supplies as a natural virtue. This is because holding repositories of scarce money in a world where the population or economy is expanding puts such a player in a powerful position, relative to, say, a student who might need to borrow money (deflation is a scary experience for any vulnerable debtor).

While many monetary conservatives actively hold power within the fiat banking system, it is politically useful for them to argue that money should be *commodity-like*, as if it were a rare substance that must be carefully rationed. This mythic tradition runs deep in conservative thought. We all have a tendency to present our political arguments as being in alignment with innate human nature, and, while I might argue that humans naturally love staring into fires and dancing in groups, conservative economists have a different take: they argue that humans are *naturally* individualistic profit-seekers whose actions lead to 'natural market order' if unimpeded, a position that in turn requires a natural monetary ideal. This role is given to gold, which functions as a contrast to the 'unnatural' money 'created from nothing' by collective institutions.

Gold was created through ancient star explosions, and this lustrous star-debris has a long history of being used as jewellery. It also has a difficult history of being seen – in certain contexts – as 'money' or 'money-like'. It was not universally seen as such, and even when it was, it was often *issued* by states and coexisted with non-commodity monies. Regardless of these nuances, it is very impractical because there is so little of it: for the 7.7 billion people on the planet to use it, it would have to be ground into a fine dust (and if we did start using such micro-fragments, those who currently hold tonnes of it would become immensely powerful). In conservative economics, then, gold is held up to be emulated rather than used, serving as a kind of platonic ideal. This was even true during the so-called

Gold Standard, when the metal was ritualistically passed around between central banks. It added extra gravity to their monetary authority while constraining their human-created institutional power with a kind of geological backstop.

The claim that Bitcoin was 'digital gold' must be seen in this context. A person entering a Bitcoin meet-up from 2008 onwards was often greeted with a monetary salvation story. It contrasted an imagined past of gold with an imagined present of corrupt fiat money, 'backed by nothing' but political alchemy. This in turn was contrasted to an imagined future, in which cryptography, backed by strong and pure mathematics, looms as saviour.

People entering these early Bitcoin meet-ups were not always monetary conservatives. There were hardcore techies, open-source software advocates and even a sprinkling of anti-capitalist anarchists and New Age gurus. Many were drawn in by the fascinating technological novelty required to achieve the decentralisation aspect of the system, but that in turn made them susceptible to accepting at face value the accompanying monetary backstory. Conservative monetary thought is fairly easy to sell to people, because it draws upon the 'money user' mentality (discussed on pages 50–2) that many of us have. It is straightforward to convince a small saver that constraining the money supply is in their interest, even if it is not, and this is how politicians like Margaret Thatcher made their austerity programmes seem like homely common sense in the 1980s. The Bitcoin system took this same austerity mindset but sought to embed it in a transnational non-state unit with an ingenious design.

A rough sketch

To understand the fascination around the decentralisation aspect of Bitcoin, ask yourself what you would do if you were tasked with building a system that enables total strangers to move digital units between themselves. How would you deal with possible frauds and 'double spending' attacks faced by banks without the centralised systems of access control that banks normally use? Who would host the accounts? Who would check balances to authorise transactions? Would they get paid salaries for this work?

This is the conundrum solved by Bitcoin's ground-breaking 'recipe'. The task of explaining the components of the recipe, and their sequencing, is tough, so let's start with a crude approximation of the end result. Picture all the commercial bank and central bank data centres I described in the first part of this book. Bank data centres are private, and banks are tasked with making alterations to the accounts within them when you request it. Now imagine by contrast a large public database, visible to anyone in the world. It is not hosted in a big datacentre, but is rather hosted in many locations at once, by a scattered global network of mercenary techno-clerks, who are tasked with making alterations to it. This is an impressionistic sketch of the basic Bitcoin infrastructure. Now we can fill in a few more details.

You have something akin to an 'account' on that open database. It is called a *public address*. This public address is a pseudonym – it is visible to anyone in the world but does not directly name you as a person – and anybody on the system can send tokens to it.

For you to move those tokens to someone else, you must compose a request to the techno-clerks. The process is analogous to composing a digital letter authorising a change, and then wrapping that in a sealed digital envelope. Before sending it must be signed, a process that involves being in possession of a private key that proves you are the rightful owner of the public address holding the tokens.

To compose and send this message to the techno-clerks, the ordinary user will have downloaded a 'wallet'. The term is somewhat misleading. Think of it rather as being a little bit like a single-purpose email client, designed for the sole purpose of sending these digital envelopes into cyberspace. The wallet broadcasts the envelope via a peer-to-peer network: picture it being passed from 'hand-to-hand' (or, more accurately, from computer to computer) across the Internet via a network of peers, rippling out.

The techno-clerks are scattered all over the world and are waiting to catch these incoming digital envelopes. Once received, they add them to a pending queue of wannabe alterations to the big public database. Their task is to verify that the requests are legitimate, after which their objective will be to update the database to instantiate the requested changes, thereby moving your tokens to someone else's public address. If they are successful in doing this, they get to reward themselves.

This description is useful but inaccurate. I've used it to convey the basic spirit of the system – and you may get by with such an understanding – but it's a metaphorical simplification. For a fuller understanding we need to wade deeper in five conceptually separate steps. The first will be to describe the concept of syncing, and the second will be to describe the concept of a 'blockchain'. I shall describe them in stand-alone terms, as two distinct concepts that do not require each other, almost like two colours on a Rubik's cube with their own logic, unaware of each other. The third step, however, entails twisting them together into an elegant combination. The final two twists will complete the picture. The description that follows is intended for a non-expert reader, and it will skip the more nuanced details of the system. But even a simplified description of Bitcoin can be intellectually demanding. For anyone who struggles, take heart in the knowledge that the going gets easier on page 204, where we shall pick up again on the monetary dynamics of the system.

1. Syncing separation into unity

Imagine an open-air theatre with a thousand audience members watching a dancer. Each watcher has a laptop computer with an Excel spreadsheet open, and each time the dancer moves they must describe it in one spreadsheet cell – for example, 'took one step to the left'. It will be hard to keep their accounts of the situation *in sync*. If the dancer moves slowly the watchers might stand a chance of keeping their records aligned, but as the dance gets frenetic they will almost certainly begin to diverge. They might record a thousand different accounts of what happened.

This is the core problem faced by a 'decentralised' digital system. The 'techno-clerks' of Bitcoin are like the watchers in my theatre metaphor. One watcher may be in Turkey, while another resides in Washington state, but they are sitting in cyberspace, like audience members in different parts of a global auditorium, trying to keep a synced-up record of token movements.

Keeping digital systems in sync is a concept many of us are familiar with, but only on a small-scale personal level. For example, I have a popular note-taking programme called Evernote installed on my computer, but also on my smartphone. I can write a note on my computer, click 'sync', and watch it pop up on my phone too. The devices send the actual notes to be stored in a central server controlled by the company, which will replicate the changes across any devices connected to it. I can continue editing the note on my phone, press 'sync', and see a new version replace the original one on my computer. If I edit the note on my computer but fail to sync it before editing the doc on my phone, it jams the system because both devices are competing to alter the central system to update the other. In this case the central system sounds an alarm and creates a *fork* in which there are suddenly two versions that have diverged. If this happens, Evernote prompts me to manually 'resolve conflicting changes' to get back in sync.

In Evernote this snag is relatively simple to resolve, because I – a single person – control both devices, so in essence I am in conflict with myself over a record stored in a central system. Now imagine, by contrast, a system in which millions of people come to rely upon unco-ordinated watchers across the world to keep track of edit requests, and to convert those into a synced-up record of changes that are not stored in a central system. That's going to be *hard*. The true beauty of distributed blockchain systems is that they have a method to do this.

Systems like Bitcoin are designed to *sync* the separate databases of separate watchers into harmony, such that these watchers – who I previously referred to as 'techno-clerks', but who are called 'miners' in the system – can keep the same account of reality, despite not knowing one another. The protocol enables them to reach *consensus*, so that if we were to superimpose their separate databases over each other there would be no divergence, as though they had a *single database*. To return to our theatre, picture the watchers syncing their separate spreadsheets, and thereby giving rise to the illusion of a single ghostly spreadsheet hovering over the centre of the stage. It has a public character, and is not subject to arbitrary individual private changes, even though it lives through their separate private computers.

2. Weaving a timeline sculpture

As useful as that 'multiple synced spreadsheets' image is, it is inaccurate in another way. The thing being synced into unity by each watcher on the Bitcoin system is in fact more complex than a spreadsheet. Each is building, and then trying to sync, something that looks far more like an intricate digital sculpture, woven together with a repetitive stitching pattern. We need a new metaphor to illustrate this second concept. The metaphor is that of *timelines*.

Today we are surrounded by timelines. You can scroll through your Facebook, Instagram or Twitter page and see a chronological list of things you have posted in a 'timeline'. Similarly, you can look back through your WhatsApp messages and see a timeline of interactions with different people. We also have a range of physical objects that could be manually arranged into a timeline, such as old baby photos placed into photo albums, and school progress reports filed consecutively in a folder.

Imagine, as a personal project, you decided to consolidate these into one mega personal timeline, starting from your birth certificate, and progressing chronologically through time, piling every message, text and photo into a massive spreadsheet containing a blow-by-blow chronological account of every communication, event and post in your entire life.

As I write this I am 13,989 days old, so if I started with my birth certificate, moved through my early baby pictures, and ended with the WhatsApp message I sent today, my spreadsheet would be hundreds of thousands of entries long. If done properly, this 'Brett-sheet' would be like a digital sculpture of my life, starting with me as a pinprick embryo radiating outwards in a timeline into my adult self. Perhaps, though, one night when feeling jaded, I wince at reading an embarrassing teenage poem I wrote, and impulsively *delete* it out of my Brett-sheet. I seal up the gap, leaving the entries before and after intact. I just erased a small section of my past.

But what if there was a rule to building a Brett-sheet that prevented this, one that turned it into something more like a *Brett-chain*? Imagine if, every time I added a record to it, I was first required to make a miniaturised thumbnail of the previous record, and to include it in the new entry I was adding. The image below shows a simplified representation of this idea. In the first entry I record that I 'Went to beach'. I then wish to add a second entry, which is 'Took drugs', but before I am allowed to enter it, I must add a thumbnail of the preceding entry to its bottom corner. Similarly, to add the next entry – 'Swam naked' – I include a thumbnail of the entry before, which itself contains a thumbnail.

This is a simplified example, but the thumbnails within thumbnails create a kind of infinite regress back to the point of inception. Every new entry will embed a memory of all entries that led up to it.

Doing this creates a new property. Imagine, for example, that a few years later I decide that it is prudent to remove that second entry, and sanitise my history by simply claiming that the progression was 'Went to beach', and then 'Swam naked'. I suddenly face a problem that I did not face in the ordinary Brett-sheet. The entry 'Swam naked' contains a trace of the entry that I am trying to remove. In its bottom corner it has that thumbnail that says 'Took drugs'. A person inspecting this chain will be able to see a *break*. Any attempt to alter a past entry will create a chain-reaction disunity up the chain. Like the film *Back to the Future*, you cannot change the past without affecting the present.

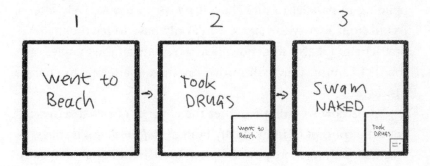

This 'Brett-chain' is fanciful, because it is recording a timeline of my life. It is, however, a great way to capture the spirit of a *block-chain*. Rather than a personally created record of my life events, a blockchain is constructed by a watcher, by catching those digital envelopes sent out by people, compiling them into bundles, and then trying to stitch them to a huge timeline, hitching each new bundle to the previous one with a little thumbnail of the bundle that preceded it. They are trying to build a digital sculpture out of requests to move digital tokens.

3. Twisting concepts 1 and 2 together: syncing the sculptures

In the Brett-chain example above, I *individually* built the chain, and if I really wanted to alter the past, I could manually change every thumbnail to reflect the change I desired (erasing the 'Took drugs' thumbnail from the 'Swam naked' block, and rewriting a new thumbnail with 'Went to beach'). If I am trying to do this a few years after the fact, it would be extremely time-consuming, because I would have a lot of thumbnails to rewrite, but I could do it.

Imagine, though, a new Brett-chain rule. Imagine that every time I add a new entry, I send the new piece to 100 friends, who will add it to back-up copies of my Brett-chain that they are slowly building in parallel to me. Now, if I wish to retroactively alter a record from a few years ago, I won't only have to rewrite my own Brett-chain: I'll also somehow have to rewrite all their copies of it too. If I cannot, they will notice my private attempt to change history.

This is how we bring together the concept of *collective syncing* with the concept of individually built *digital timeline sculptures*, to create something more powerful than both. In our original theatre metaphor, the watchers were getting consensus on a mere spreadsheet, but the true soul of crypto systems is that the watchers are getting consensus on an intricate chained timeline that each is building in parallel. Indeed, it is confusing that the term 'blockchain' is often presented as a singular noun, because it should be *plural*. Blockchain systems are *distributed,* but this does not mean 'one digital sculpture split into parts held in different places'. It means 'many different sculptures built in parallel in different places, synced into unity'. Through this syncing, the illusion is achieved of a singular, living sculpture of token issuance and transfer, one that cannot be changed once that consensus is reached (at least in theory).

4. The penultimate twist

We need one more alteration to this picture. In our metaphor the dancer first dances, after which the watchers try to capture it in their timeline sculptures. A system like Bitcoin, though, reverses this ordering, because – in any digital token system – it is the act of recording a change that *makes it happen*. Imagine our dancer standing static like a mannequin, but being able to shout requests to the watchers, and being suddenly moved, as if possessed, when they reach consensus about what he has asked. The mannequin shouts, 'Can I do a twirl?' The watchers seek to weave this into their new timeline sculptures, and upon success shout back in unison, 'He does a twirl!', which sends him spinning around.

This takes us back to our 'envelope' starting point. In the Bitcoin system, the ordinary user is a bit like that dancer, writing out a request in a digital envelope, and sending it through the peer-to-peer network to the watchers. After you have sent the request nothing changes to begin with. Your address sits frozen in the same position as you wait for the watchers to stitch your proposed alteration into their chains. And then, suddenly, you see your wish animated into reality. Your tokens move.

5. The final twist

So far the spectre of Bitcoin appears as a system for building and syncing separate digital sculptures, run by separate techno-clerk watchers, into unity, in order to animate a trail of token movements between people. A question that remains is why these watchers would take the time to do this. I characterised Bitcoin as a two-pronged attack on the banking system. So far my description points only to its first prong – its decentralised nature – but the final piece of the puzzle will take us to that second prong: the promotion of an unchangeable token supply.

In the banking system *money creation* is conceptually separate from *money movement*. As we saw on p.69–72, a bank issues chips when, for example, its loan department decides to extend credit. Those chips can then be reassigned between accounts through its payments department, which is a separate division in the bank with separate employees. In Bitcoin, however, token creation and movement are fused into a single operation. This is because the system has a rule that can be summarised as follows: if a techno-clerk is successfully able to collect up the incoming digital envelopes containing requests to move existing tokens, and to update their digital sculpture with them, they earn the right to *write a set quantity of new tokens into existence for themselves* and add it to their sculpture (originally this quantity was fifty, but it reduces over time). If everyone in the system reaches consensus to accept their new updated sculpture, then that techno-clerk will find themselves in possession of new tokens.

This creates a new problem. In creating that rule, the system simultaneously creates an incentive for the techno-clerks to update their sculptures as often as possible, which means the units they would get as a reward would proliferate wildly. To act as a counter-weight to this, the Bitcoin system makes it deliberately hard for them to update their sculptures. This is where Bitcoin's now infamous 'proof-of-work' system comes in.

Back when I still used Gmail, I enabled its (now discontinued) 'drunk filter'. As I clicked 'Send', a maths puzzle would pop up on screen, which had to be solved before the email would send. The rationale for this quirky feature was to prevent drunk people sending a message they would later regret, by *slowing them down* and thereby giving them time to re-consider and cancel it. The maths puzzle was not very hard – it added maybe ten seconds to the process of sending one email. Imagine, however, if I was trying to spam 10,000 people: the drunk filter would force me into about twenty-six hours of calculation. Confronting someone with a 'proof-of-work' obstacle like this is a great way to prevent spam.

The Bitcoin proof-of-work system is akin to a very difficult obstacle course that pops up before any techno-clerk can show the world their new sculpture. Bitcoin uses the SHA-256 hashing algorithm for this, and it creates a puzzle a lot harder than the Gmail drunk filter. In fact, the only way to solve it is to use 'brute force', trying thousands of random answers in the hope that it works. A side-effect of this is that the techno-clerks in the system have to expend a lot of energy. In the Bitcoin system, this is all automated – they run 'mining rigs' (big, computing-power-heavy server racks) that automatically run through all the steps. If we had to use a human metaphor, though, imagine them, having collected a bundle of digital envelopes to add to their sculpture, all sweating while trying to press 'Update', as if there were a force field repelling them. Eventually one techno-clerk breaks through, and, shouting in victory, sends out an updated version of their sculpture (along with a set number of new tokens built into it for themselves) to everyone else, who must abandon their own attempts, sync up, and begin the process anew.

By slowing down the updating, the ability to create new units as a reward in the process is slowed to a trickle. It is this mechanism that makes Bitcoin tokens feel like 'objects' wrestled – or 'mined' – out of cyberspace. The arduous act of recording both creates new Bitcoin tokens and animates previously created ones to snake forward in long chains of recorded action.

This is why early Bitcoin enthusiasts wore shirts with slogans like 'In Mathematics we Trust'. In much the same way as a language is brought to life by those who speak it, while resisting the control of any one speaker, the Bitcoin system is brought to life by a network of people who use the common protocol. Once activated it feels as if it exists outside of the control of any one of those individuals. This might appeal to those who feel spooked, betrayed or disillusioned by large institutions that seemingly have the ability to alter the fabric of reality. It explains our early Bitcoin meet-ups, which made reference to unbroken chains of action throughout time. This new 'crypto-leviathan' ties up people in an unbreakable honour code, such that honour is not required. All past actions are tattooed onto the present. Everything has a lineage. Nothing is just 'created by fiat'.

Cult of the blank token

There are many philosophical and political questions that such a design might provoke. For our purposes, however, we need only ask one: what exactly *are* the tokens created and moved around in this elaborate process? The original Bitcoin Whitepaper was sub-titled 'A peer-to-peer digital cash system', so Nakamoto believed the tokens were 'digital cash', and the term 'cryptocurrency' was quickly applied. I was an early user of Bitcoin, but felt the monetary language to be premature. Much like an obscure director might call their unrecognised movie *The Greatest Show on Earth* (in the hope that it becomes that), enthusiasts were posturing when applying monetary language to the tokens. They were looking forward to a time when the tokens *would* become money, but much media reporting went along with this present-tense branding of a future state, with reporters and social media influencers referring to the tokens as a 'new digital currency'.

This returns us to the point I began the chapter with. We are susceptible to seeing movable numbered objects as 'money', because we fixate upon the surface-level numbers that accompany a monetary system. Some numbered objects are easy to see through: a chocolate coin moulded with the number '1' and packaged in a foil wrapper with money-like branding. We understand that the branding is just a light-hearted coating over a chocolate. But what happens if you are presented with a numbered *digital* object wrapped in money-like branding? There is no physical body, so all you see is the branding and numbers.

Something curious occurs, though, when you ignore the branding, and stare only at the numbers carved into Bitcoin's timeline sculptures. For example, imagine you see this.

1

What is this number? Does it point to something? If this was rendered on a screen it might look a little bit like a number you might see in a bank account, but a bank account is called 'account' for a reason. The numbers in a bank account are not free-standing mathematical objects: they are *accounting records* of legally backed IOUs issued out to you by the bank. Thus, when a bank 'writes numbers' into your account, it is an act of them granting you access, and that act legally *puts them on the hook*. Those numbers are a *liability* to them, not something that they have created and given to themselves as an asset (much as a voucher issued out by a coffee-shop is a *liability* to the issuer). The only way the numbers could be separated from the IOUs they represent is if some extreme event like an apocalypse destroyed the bank and the entire legal system, thereby rendering the once voucher-like numbers in your account as *mere numbers* with no power in the world.

In the Bitcoin system, however, numbers mean something different. They might have the surface appearance of something in an account, but the techno-clerks who write them out are not *accounting* for anything, and the number certainly is not a liability to them. No, they write them out after expending energy – just as I might exert a lot of energy to engrave the number '1' on a bottle top and then say, 'Here is my currency'. The key difference, however, is that a bottle top is a physical object that can be distinguished from the number I engrave it with, whereas in the Bitcoin system there is no physical object. The object and the number are one and the same thing.

Put another way, Bitcoin 'tokens' are just written-out numbers, recorded on the blockchain system, somewhat like an accounting system with nothing being accounted for. The system places limits upon how many numbers can be written out, and it allows the participants to perform *operations* on them: thus, once created, a 50 might be fragmented into fractions that can split off in different directions that end up in different addresses in the Bitcoin system.

The key nuance, however, is that the community around Bitcoin has created a separate *psychological layer* that has been pasted over the underlying reality of the system. Rather than seeing themselves as moving limited-supply numbers around, the system participants are encouraged to imagine themselves moving a type of 'digital commodity'. The Bitcoin community tries hard to associate the numbers with metal, relying heavily on the visual imagery of Bitcoin's *branding* to do this. The numbers come wrapped in pre-packaged imagery of futuristic metallic 'coins', an image that is reinforced by the choice of brand-name – Bit*coin*. This is why, in the early days of Bitcoin, photojournalists desperate for a physical image to place in news stories flocked to take photos of a metal trinket with the Bitcoin logo on it.

Perhaps most importantly, however, a highly vocal community of promoters exert much effort in creating a *story* about this imagined 'coin'. The tokens might solely be numbers, but from a certain angle they come 'from effort': the energy-intensive proof-of-work mechanism can be tied to the imagery of a gold mine (which slowly outputs gold into the world), and it can be pitched as an alternative to the idea of states as sorcerers creating 'money from nothing'. If it takes a great effort to write out '50', and you are only allowed to do it a set number of times, a *mental image* of 'digital gold' can be conjured, but it remains an unstable image. Imagine asking an Olympic athlete if they would like to swap their actual gold medal for a 'digital gold' medal. Even if they choose to buy into the mythology around Bitcoin's numbers, they are going to have a hard time – as tactile beings – trying to feel excited about this imaginary object.

Thus, while gold and fiat IOUs are, unlike Bitcoin, more than *mere numbers*, Bitcoin still tries very hard to associate itself with the former by contrasting itself with the latter. But this raises a new question. Isn't modern gold something you buy for money, and possibly re-sell for money, rather than *being money*? Can a digital token based upon (tenuous) gold imagery really serve as an alternative monetary system? Won't it just become a collectible with a monetary price, like gold?

207

Countertrading collectibles

A national economy is a giant interdependent network structure in which diverse goods produced by diverse people all route through a common money system. We find it difficult to see that structure, but we get a glimpse of it in the aisles of a supermarket. A supermarket has diverse goods, produced by diverse companies, but they all carry price stickers with the same symbol, all pointing to a common monetary web.

From a young age, a child in a supermarket can see that *money* is not the same as *everything priced in money*, and that the key hallmark of the former is that it makes them think of the latter. If given a £20 note for pocket money, a nine-year-old might project their mind towards sweets on shelves, or a small toy. Likewise, the child's mother looking at her bank balance is projecting her mind towards commodities like gas or porridge, or services like childcare. Conversely, if she looks at individual goods, she either thinks of them in terms of a specific utility ('I use my car'), or in terms of money ('I wonder how much I can sell it for?'). She does not think of her car in terms of porridge. Likewise, she does not think of money in terms of other money, unless she is crossing a geographic boundary from the British pound system into, say, the South African rand system, within which supermarkets price porridge using a different symbol.

There are some obvious features about Bitcoin tokens that allow us to categorise them. First, there is no supermarket in the world in which porridge is priced in Bitcoin tokens.* This means Bitcoin tokens are not participants within the *foreign exchange* market. Secondly, Bitcoin tokens have come to have a *price*, denoting the quantity of money someone will outlay to obtain them (indeed,

---

* The situation in El Salvador is explained on page 210

there is a very large amount of trading, and Bitcoin's price has been a favourite topic discussed in the media and social media over the years). Given that these tokens are not on the foreign exchange market, this must mean they are goods, not dissimilar to items found in a digital supermarket. This is confirmed by the fact that – upon holding such a token – people generally do not think of other goods, but do imagine a *resale price* in money.

Thirdly, given the commodity imagery around Bitcoin, the tokens have come to have the feel of limited-edition *collectibles*, albeit ones that cannot be displayed like a gold amulet can. Fourthly, given that they have no visible body, and can be transferred digitally, they are far more *movable* than the average item you might find in a supermarket. Think of Bitcoin tokens as being movable numbered objects, wrapped in monetary branding, which have been turned into collectibles with a monetary price. These do not directly challenge the monetary system, but certainly do piggyback on its imagery, and integrate with it through a phenomenon called *countertrade*.

Countertrade is the process of swapping money-priced goods. Imagine I wish to return an item to a supermarket: I can either ask for a refund – they give me my money back – or I can ask to *swap* it for something of equivalent price. In this situation you might see me handing over the old thing and getting the new thing, and that act – superficially – might look like I am 'buying' the new thing with the returned good. That, though, is an illusion. After all, I could have asked for the refund money from the cashier, and then passed it back to them to buy the new thing, but what I have done instead is blend those into a single action by cancelling out the two offsetting money movements. This process of 'clearing' money-denominated objects against each other goes by many names in the world (such as 'calling it quits'), but the technical term is 'countertrade'.

When we imagine an entire economy as being like one vast

'supermarket', countertrade appears in a more complex form. A person holds a good, and, rather than reselling it and using the proceeds to buy something new, they swap it, guided by the two monetary prices. All acts of 'buying' things with Bitcoin are acts of countertrade. Someone takes a movable digital collectible and swaps it for something of equivalent price. The best place to see this in action right now is in the dollarised country of El Salvador, where a Bitcoin-enthusiastic president pushed through legislation in 2021 forcing merchants to accept the crypto-token. This has led some people to believe that goods are now priced in Bitcoin in El Salvadorian stores. In reality, the ever-fluctuating 'Bitcoin price' for a box of porridge in El Salvador is actually a US dollar price, but one that is refracted through Bitcoin via a countertrade ratio that constantly changes as the dollar price for Bitcoin changes. In the time it takes you to finish an El Salvadorian restaurant dinner, the so-called 'Bitcoin price' of your bill could have changed forty times, while the real dollar price of your meal stays constant. But, rather as a mirror imparts an image of the thing it reflects, while its glass remains invisible to you, an invisible collectible with a dollar price and money-like branding can easily feel 'moneylike', leading to the person imagining themselves to be using it to 'buy' the dinner (when in reality it is the dollar price that is truly doing the buying).

When you swap supermarket goods, you are not fighting the monetary system. Similarly, when you countertrade cyber-collectibles, you are not bringing down the global monetary order. Nevertheless, if Bitcoin were to greatly increase the use of countertrade – a practice that is historically limited – it could certainly add new texture to the surface of our existing monetary system, and mess with the standard paths used. In Chapter 6 we saw that new underground forms of payment might be inspired in the wake of an excessive attempt to clamp down on cash. Bitcoin's high 'countertradability' certainly makes it a candidate for that role, despite its volatility.

12

The Political Tribes of Cyber-Kowloon

In 2015 I was interviewed for a Dutch documentary called *The Bitcoin Gospel*, which chronicled the rising cryptocurrency movement. The crew filmed me on a boat on the Thames as we cruised past the huge banking skyscrapers of London's Canary Wharf. They had wanted to film me standing underneath those buildings, but the entire area is a privately owned estate with its own security force, and the estate would not grant permission to the film crew.

Against this backdrop the filmmakers asked me whether Bitcoin was like a 'new Occupy Wall Street', leading the charge for a new open financial system. After all, its rise to public prominence came after the decline of the Occupy movement in 2012. I answered 'Yes', but I was doubtful about my assessment. Occupiers were a diverse group who believed the state had been captured by corporate interests, and that it must be wrested back by a public working together to make a more equal society. Bitcoin promoters, by contrast, were overwhelmingly male and far more intolerant of government than of corporations. Many distrusted the concept of 'society', preferring to imagine a world comprised of competing individuals.

The Bitcoin scene was – to put it bluntly – more right-wing than

Occupy, and heavily influenced by pro-market libertarianism. The most hard-line free-market libertarians see capitalism and its markets almost as a type of deity, offering rewards to some and punishment to others. Libertarian Bitcoiners tend to portray the existing financial system as akin to a corrupt Catholicism, in which big institutions – governments and banks – mediate access to holy profit while distorting the will of the market. In this scenario, they are like the Protestant Reformation, seeking direct communion with the market. To the left-wingers of Occupy, who are sceptical about profit, such reformers might look like corporate capitalists under a different name, given that both claim to revere market processes. But that is like saying Martin Luther and the Pope have more in common than their differences.

I use the religious imagery deliberately, because there is a strong strain of millenarianism in Bitcoin circles – the belief that a fundamental restructuring of the world is imminent. Roger Ver, a libertarian entrepreneur who was featured in the Dutch documentary with me, even earned the nickname 'Bitcoin Jesus' for the fervour of his belief. Ver's passionate speeches about Bitcoin hinted at a story of 'manifest destiny', in which those who devoted themselves to bringing forth the new crypto-world would be rewarded, leaving the old world behind.

This same message can even be found in Bitcoin's numerical setup. The original coding stipulated that the techno-clerks could only ever bring to life 21 million tokens (the number is arbitrary, given that it could also have been 50,000 or 100 million tokens). If those were equally spread among the 7.5 billion people in the world, each would get 0.0027, but the early system gave them out in 50-token chunks (18,333 times the egalitarian distribution) to whichever techno-clerk had the right equipment at the right time. In 2018 Elon Musk earned 18,333 times the amount of someone earning $28,000, but he did have to launch a global corporation to get into that position. Early Bitcoin miners, by contrast, claimed disproportionate percentages of the tokens for little more than letting their big computers run.

The comparison above is perhaps misleading. As we've seen, Bitcoin tokens are closer to limited-edition collectibles, and collectibles were never designed to promote egalitarianism. Their psychological structure depends on a sense that they are hard to get hold of. Nevertheless, these collectibles had monetary branding, which suggested they were *supposed* to be money. But as speculative profits from buying and selling them rose, many Bitcoiners (a name given to enthusiastic promoters of Bitcoin) gave up on the idea of the tokens as 'money', and started seeing them as investments to be bought with the intention of resale for capital gains. This created much confusion (which continues to this day) because the original monetary language began to clash with a language of investment: thus, a rise in the price of the collectibles was called 'deflation' by some, and 'dollar gains' by others (for context, imagine holding rare art that rises in price, and then referring to that as 'deflation'). Others ignored the cognitive dissonance by using both types of language simultaneously, but it was obvious that the tokens were being absorbed as goods to be traded within the standard monetary system.

The idea that Bitcoin was 'digital cash' nevertheless clung to the collectibles, a label that began to cause problems for my real-world campaigning around physical cash. Every time I published an article about the need to protect the cash system, it was overrun with comments from crypto hopefuls claiming that Bitcoin solved the problem. The banking and fintech sector was trampling unimpeded over the physical cash system, while these rebels placed their hopes in a cyber-token that nobody used for supermarket commerce (although, as touched on in the previous chapter, they could be used for countertrade). Pointing this out, however, would draw out the millenarian story: *they will become money in future*.

The speculative trading in Bitcoin collectibles resulted in a massive increase in people trying to send transaction requests back and forth in the system, which meant it started getting clogged. For those who still held onto the idea of Bitcoin serving as money, this did not bode well. The Bitcoin protocol has a group of core developers that make updates to it, so Roger Ver – Bitcoin Jesus – demanded that they update it to increase the number of transaction requests that could be processed. The core developers refused, so Roger left with his followers in 2017, 'forking' off to create a rival version called Bitcoin Cash, thus transforming him into Bitcoin Cash Judas.

'Forking' is a core practice within open-source culture. Imagine yourself as a teenager with different possible futures ahead of you, and – in a state of extreme indecision about which to follow – somehow splitting into two separate adult versions of yourself. Each shares the same childhood, but follows a different path into the future. This is how crypto forking happens. Bitcoin Cash shares the same underlying code as Bitcoin and spent its formative years as one with the Bitcoin blockchain, but split away from that shared past to create a parallel reality under a new brand name.

Strange allies in the war on cash

Roger Ver and his followers began to argue that Bitcoin Cash is the 'true' Bitcoin, somewhat like a breakaway sect claiming to maintain the spirit of a heretical prophet while berating the original sect for losing its way. Early Bitcoin had equated itself with digital gold, digital cash, commerce and saving, but the divided community fought over those labels. The Bitcoin community claimed Digital Gold and Saving ('Bitcoin is digital gold for hoarding'), while Bitcoin Cash claimed Cash and Commerce ('Bitcoin Cash is digital cash for commerce').

Physical cash – as we saw in Chapter 3 – is the only government money we can hold, our normal digital money being privately issued bank chips. This can put libertarians in a quandary over physical cash: they appreciate its anonymity while disliking its issuer. The prospect of creating a non-state 'digital cash' is therefore a hot topic, and my work on this got me invited to the first major Bitcoin Cash global gathering – an event called Satoshi's Vision hosted by Ver in Tokyo. I found myself something of an ideological outsider, though. My position on cash comes from a critique of financial capitalism, but libertarians are prone to framing the same concerns about surveillance and control as a critique of 'financial socialism'. Upon the walls of the toilets were stencilled phrases like 'Who is John Galt?' – a reference to Ayn Rand's *Atlas Shrugged*, a must-read for conservative libertarians. In the book Galt is a brilliant entrepreneurial inventor who refuses to yield his designs to authorities, and Rand presents him as a radiant force standing in opposition to collectivist socialist tyranny.

The horseshoe theory of politics asserts that those on the fringes of both the political left and right can end up having more in common with each other than those in the centre. This partly explains the tentative answer I gave to the documentary filmmakers when asked about Bitcoin and Occupy. Many Occupiers drew heavily upon left-wing anarchism, which itself draws heavily upon *pre-state* and pre-capitalist imagery of highly communal groups (think of anarchist community gardening projects, tool-sharing initiatives and breakaway communes). That crypto-tokens diverge from the political centre draws the attention of these left-wing anti-statists, who occasionally turn up to crypto meet-ups to share space with pro-market libertarians. The latter, however, channel a *post-state* mentality. I described in Chapter 10 how state leviathans break down the requirement for communal solidarity by enabling strangers to interact. In the same way that parents might give life to teenagers who resent

215

them, these leviathans can catalyse an individualistic world view, which then turns its resentment towards them. The anti-statism of crypto libertarians is framed as a desire for a world of individual 'do it yourself' entrepreneurs, rather than a desire for tightly knit communal gardening projects.

Like a teenager fantasising about running away from home, the libertarian imagination generates images of markets untethered from a state foundation, and some libertarians will even act out these fantasies. The US Bitcoin trader Chad Elwartowski, for example, rigged up a seasteading dwelling fourteen miles off the coast of Thailand before being forced to flee after the Thai government accused him of violating its sovereignty. On his Facebook page he posted, 'I was free for a moment – probably the freest person in the world.' Seasteading was a theme at Satoshi's Vision too, with a Bitcoin Cash promoter proudly showing me pictures of offshore living pods he was designing. He went on to tell me the story of Kowloon Walled City, a formerly ungoverned space on the outskirts of Hong Kong. In the absence of regulation, Kowloon's buildings crammed together like overgrown concrete vines, creating a labyrinthine maze. It was demolished in 1994, but he saw it as a symbol of unregulated human endeavour on the periphery of governed zones.

These are the images lingering in the background when crypto hopefuls imagine a 'land of the free' in cyberspace. Could a 'cyber-Kowloon' be built, a parallel digital reality that you can slip into? The libertarian activist Cody Wilson frames Bitcoin exactly like this, when proclaiming that 'Bitcoin is what they fear it is, a way to leave . . . to make a choice. There's a system approaching perfection, just in time for our disappearance, so, let there be dark.'

Cody is drawing upon 'dark market' imagery, associated with the dark web, and made this statement while promoting a (now defunct) Bitcoin project called Dark Wallet. The cypherpunk goal

Kowloon Walled City

(discussed on pages 183–4) was to co-opt Internet infrastructures to build shadow networks, and many crypto promoters draw upon this same 'dark' imagery, gravitating towards visions of being under-ground rebels (even as they split into rival factions such as our Bitcoin Cash community). This shadow cyber-economy is one part real – crypto tokens do get countertraded in actual dark markets – but one part mythological: the story adds intrigue and excitement to tokens that are, for the most part, treated as collectibles by peo-ple living lives within mainstream corporate capitalist society.

This latter element perhaps explains why the community is heav-ily male. The imagery of rugged heroic individualism alongside trading is enticing to men who might otherwise be working a stand-ard day job. Launching digital collectibles that tap into this can be very lucrative, which means the factionalism is now as much com-mercial as it is political. Since Bitcoin, hundreds of 'alt-coins' have emerged, Bitcoin-like clones with new branding. These tokens – such as Litecoin, Peercoin or Dogecoin – maintain the same basic collectibles structure by issuing a limited edition set of numbered tokens, but tinker with their logo, quantity, mechanism or privacy. Cyberspace is now littered with countless crypto-tokens sold in exchange for US dollars. Many of these first-wave systems should be seen as proof-of-concept initiatives (albeit ones that enabled their promoters to become extremely wealthy), but they have paved the way for more sophisticated iterations, to which we will now turn.

Cyber-Kowloon gets sophisticated

Imagine children rummaging through a box of coloured plastic tokens. For a while the variety of colours provides interest, but eventually the children seek to enchant the tokens through imaginative leaps like 'These are tanks on a battlefield!' One of the first steps towards higher sophistication in crypto communities looked somewhat like this, as innovators began to wonder whether crypto-tokens could be fused into things in the real world. It was a desire to move away from numbers as nouns and towards the numbers as adjectives, creating tokens that promised access to *things beyond themselves*. A generic token could be made specific by, for example, transforming it into a voucher for physical commodities, or a share certificate promising future returns.

While our kids might create an imaginative connection between a token and a tank by simply stating that it exists, the adult world requires something more robust to forge an enduring link. The mere act of saying, 'This token represents one tonne of platinum,' is a tenuous assertion unless the law courts will recognise the claim. In the absence of this, a hard-coded link is required. To give an example, consider a *key* to a locked storage room full of platinum.

218

The key itself does not store anything, but is tied to the platinum through the fact that the metal cannot be accessed without it. Transferring the key transfers that access, and so one might call it a 'platinum-backed' key. The new challenge taken on by crypto-engineers was to turn their tokens into something like digital access keys to things in the real world.

To have any hope of providing a decentralised alternative to mainstream banking, though, crypto would need to go beyond one-way token-transfer systems. A deal involves two parties having to fulfil their side of the bargain, and this is often where our traditional leviathans thrive. If I run off with goods before paying a shopkeeper, they send the police after me, and even in the old Kowloon Walled City you might find yourself shot by gang bosses if you harmed someone under their protection. Similarly, Internet leviathans like Amazon have arbitration systems to ensure deals go through. But in the crypto realm there are no crypto-police (or gangs) to turn to if someone does not fulfil their side of the deal when you transfer tokens. Crypto systems need a way to deal with multi-step processes like 'send tokens and receive goods' or 'send tokens if work is completed', or 'grant platinum tokens to person who sends money tokens'.

By far the most captivating vision for this was offered by another pretender to Bitcoin's crown. I first encountered the team behind Ethereum in an upmarket flat in London in late 2014, six months before the launch of the network. Two of its engineers were pondering mathematical equations scrawled over whiteboards around the dining room, while one casually spoke of governments as 'outdated operating systems'. Blockchain technology can engage both the practical problem-solving mentality of an engineer, and the yearning for a political purpose. Crypto engineers like those building Ethereum were beginning to see the world as a huge social machine jammed by faulty political parts and misaligned economic incentives. With the right set of coded contracts, calibrated with just the right rewards, a sophisticated 'cyber-Kowloon' could emerge. They did not want a grubby dark web enclave. Rather, they sought an orderly utopia guided by 'crypto-economics'. The latter uses game theory – the economic study of individual incentives – to try design systems that are unprofitable to break or misbehave within.

Ethereum's key hope lay in *digital vending machines*. One does not see vending machines running away shouting 'So long, sucker!' after you insert coins. They are mechanically programmed to activate and abide by a market contract when you fulfil your part of the deal. In a lawless world, a shop might be robbed, but an armoured vending machine can still do business. The Ethereum system's key innovation was to allow people to code and deploy the equivalent of armoured digital vending machines on their network, and to give those machines their own addresses so that they could act as agents doing business with humans on the system.

In the Ethereum system, these go by the confusing name of 'smart contracts', a term first used in 1994 by the cryptographer Nick Szabo, who also used this metaphor of a vending machine to describe the concept. If a normal vending machine is made from mechanical parts, a digital vending machine is written out in code. Ethereum has a token called *ether*, which can be used to activate those digital machines. To picture this, think of a theme park that only accepts the tokens issued by the park management. The Ethereum network is like a digital theme park with installations calibrated to accept the native ether token. Just as you can code a soft-drink machine with instructions like 'If £1 placed in your slot, then release Coca-Cola,' you can code these digital machines to say, 'If 1 ether token received in your crypto address, then release 5 share tokens back to address that sent ether tokens.'

Ethereum, like Bitcoin, has a network of techno-clerks who receive requests from those with addresses (including the vending

machines), and animate them into reality. This is a more sophisticated process, because – upon activation – many of those digital
machines are required to do acts of computation. They are like
small programmes waiting to be activated on the network.

The Ethereum team – which was led from the start by an other-
worldly Russian-Canadian programmer called Vitalik Buterin –
originally raised substantial funds by 'pre-selling' these ether
tokens – like selling theme park tokens for an unbuilt theme park –
and used the proceeds to hire people to build the base infrastructure,
which they launched in 2015. The new system was like a blank slate
upon which people could project their visions of a future alternative cyber-economy. Enthusiasts imagined assemblages of smart
contracts connected together to create more complex decentralised
autonomous organisations (DAOs). These DAOs, in turn, could
become alternatives to the platforms of Silicon Valley, triggered into
action by ether tokens sent by citizens of cyberspace. Indeed, many
of those creepy ideas from Silicon Valley found their way into these
circles. Some imagined cars that could not start unless one purchased digital keys from a digital vending machine. Perhaps the car
could be remotely stopped by a signal sent from cyberspace (just as
an old public telephone would cut you off as you ran out of credit).
Others imagined fleets of autonomous vehicles roaming the highways, hiring themselves out for cyber-tokens via DAOs.

There were more down-to-earth visions too. Blockchain platforms
are said to be 'trustless', which is intended to mean that people do not
need to be trusted for the system to work. For a techie this is a practical issue rather than an ideological one: even if they believe 99 per
cent of humanity is fundamentally good, in a faceless Internet network of 10 million people it might take only *one* malicious actor to
bring the thing down. Designing systems that will stand up regardless of dodgy or incompetent actors appeals to them. That same
vision also appeals to development professionals working in tough

environments where a decentralised infrastructure might just be more resilient than a centralised one. And so people in humanitarian aid circles began exploring blockchain. I provided input into investigations run by UNOCHA (United Nations Office for the Co-ordination of Humanitarian Affairs), UNRISD (United Nations Research Institute for Social Development), Amnesty International, and the UN Environment Program.

Soon almost every major NGO was taking an interest, with climate change groups running blockchain hackathons while aid groups wondered whether the technology could be used for the distribution of food vouchers. The proposed uses ranged from tracking the movements of goods through supply chains to combatting blood diamonds and recording carbon credits. These types of groups were not interested in the fringe political philosophies that circulated in the core crypto circles. They were practical political centrists looking for new ways to do their jobs. The horseshoe was going full circle.

The finance vending machine

By 2016 the Ethereum community was split between those who held and speculated in its idle tokens, and those entrepreneurial techies or researchers who tinkered with hypothetical projects that those tokens might be used for. This bifurcation between idle token holders and active entrepreneurs created a perfect opportunity for *finance*. In May 2016 a new initiative called 'The DAO' was launched as a digital finance vending machine on the Ethereum network. It was designed to gather ether tokens from investors, pool them together and grant the contributors voting rights to determine which entrepreneurs should be funded (almost like a decentralised mutual fund). It was a confusing title, because DAO is a broad concept in crypto circles, and calling an initiative 'The DAO' is like an investment fund calling itself The Company (rather than, say, BlackRock). Nevertheless, the founders enlisted Vitalik Buterin and other leading lights in the Ethereum community to join their board. The hype worked and the entity gathered ether tokens worth over $100 million from over 10,000 contributors in just fifteen days.

The architects of The DAO did not hold back on hubris, claiming that their creation existed 'simultaneously nowhere and everywhere' and that it operated 'solely with the steadfast iron will of unstoppable code'. The first statement is a reference to the dark market realm, but to understand the second it is important to note that coded smart-contract systems, designed by crypto-engineers with crypto-economics, drew their authority from a claim that 'code is law' (precisely because they had no law courts to fall back on).

It's common to make a distinction between what is legal and what is possible, with law enforcement agencies seeking to keep possible actions within the boundaries of what is legal. Consider a term like 'speed limit', though. It has both a legal and a physical meaning. The legal speed limit might be 70 mph, but the physical speed limit of the world's fastest car might be 315 mph. It might be against the law of humans to break the first, but it is against the law of nature to break the second. Crypto systems drew on this latter imagery, claiming to be governed by 'unbreakable' forces. There is no need for human police because unstoppable code was like physics, acting as both law and police in itself.

A month later I saw this physics break down. Standing behind a prominent Ethereum developer in a Brussels restaurant, I watched him sketch a diagram on a piece of scrap paper. He was detailing an emergency plan to change Ethereum's code in order to fix a major hack that had just occurred. A rogue hacker had worked out how to

223

confuse the finance vending machine into giving them tens of millions of dollars' worth of ether tokens out of it. If the offender got away with this, they would control a large chunk of stolen tokens, but if the Ethereum developers tried to intervene, it would shatter the illusion of their system being governed purely by unstoppable code. Nevertheless, the thousands of investors in The DAO had just seen their investments reduced to zero, and many of them were not in the mood to be told that 'code is law', and that the hacker – having skilfully manipulated the code to take all their tokens – was thus a legal player. They sensed that the 'letter of the law' (the code) was not serving the 'spirit of the law', and that the former needed to be brought into line with the latter. This, however, ran contrary to the political ideology of crypto platforms, which were set up to escape corrupt and meddling politicians – the very people traditionally leading such charges to change laws.

In the final analysis, the apparently unstoppable code gave way to the real world of immovable human politics. The Ethereum team released an update to the system designed to eliminate a slice out of the Ethereum history, to make it as if the hack had never happened. To 'go back in time' like this required convincing the techno-clerks to accept the change. It was an exercise in decentralised cyberstatecraft, but it also created a band of rebels who forked off to create a new version called Ethereum Classic in which the hack still happened and in which the unstoppable forces of code still prevail. Proponents of the two now have sectarian fights, much like those seen between the different forks of Bitcoin. I tell this tale, though, because it knocked much of the naivety out of the blockchain scene, and also launched research into the field of *decentralised governance*, which considers how disparate groups of people might make decisions without an overt central authority. This is now an active field with much experimentation.

A political-economic melting pot

The stories I have recounted come from the inner circles of the blockchain movement, but by now most members of the public have been exposed to the various crazes that ripple through crypto markets and into the media, and back into crypto markets from there. There was, for example, the ICO – or 'initial coin offering' – craze in 2017. The term was a play on the phrase 'initial public offering', a finance term used to describe the process by which companies raise money by issuing shares on the stock-market. In the case of ICOs, opportunistic crypto-entrepreneurs raised huge amounts by setting up a rash of digital vending machines that spat out tokens to use in future systems not yet built. It turned Ethereum into a lawless stock market for unenforceable tokens masquerading as shares. I know many people who raised multiple millions for things that never materialised. The typical process was to build the hype, raise the money, work on the project for a year, and then just quietly slip away (a process accompanied by the gradual slowdown, and then silence, of a once-active Twitter feed).

In between the frauds and the scandals, though, there has been an undeniable movement building. Bitcoin and Ethereum are now viewed as incumbents to be improved upon, and an anarchic efflor-escence of crypto technologies have emerged, accompanied by a proliferation of political visions. In many ways these platforms have offered homes for those on political extremes. For example, in *The Politics of Bitcoin: Software as Right-Wing Extremism* Professor David Golumbia argues that the technology combines libertarian eco-nomics with far-right populism. Indeed, the former Trump adviser Steve Bannon believes in Bitcoin as a means to drive a 'global popu-list revolt'. The underbelly of the scene is also rife with anti-Semitism and anti-feminism. At a Vienna crypto gathering I met an outright neo-Nazi, who openly advocated for eugenics while a group of men listened, and remember fighting a core Bitcoin Cash developer who was insistent on his racist views. On the far fringes the alt-right 'neo-reactionary' theorist and computer scientist Curtis Yarvin started a project called Urbit as a crypto-operating system.

Some on the far-right bend the Singularity story (from Chapter 9), via blockchain, into a nihilistic Goth vision of a machinic system that will engulf everyone (like in *The Matrix*). On the lighter side, there are softer and more biological Blockchain-as-Gaia visions being taken forward by neo-shamanic and psychedelics communities. The Assemblage in New York is an elite workspace where well-spoken New Age social entrepreneurs discuss consciousness and spirituality over a vegan buffet served daily. Here I watched a financial trader sit in a yoga pose, in ethnic dress, speaking with a blissful smile of the possibility of Blockchain as a spiritual financial revolution. In keeping with horseshoe politics, as the far-right blends in ideas associated with hippies those poles meet, and vice versa. Peter Thiel (of PayPal and Palantir) has a fund that invests in Urbit, psilocybin companies and a Bitcoin mining company that claims that Bitcoin will help solve climate change through its huge energy consumption. Go figure.

There are many tribes imagining their own version of cyber-Kowloon, including one that has escaped our attention so far. If there is one thing that attracts the interest of the banking sector, it's when people get rich, and bankers noticed that crypto-entrepreneurs had managed to make a *lot* of money – normal money – by tapping into the idealistic imaginations of different communities. Perhaps they could also get in on the business of selling visions of cyber-Kowloon?

13

Raiding the Raiders

Astana (Nur-Sultan) rises out of the Kazakh Steppe, the immense stretch of grassland in northern Kazakhstan. The steppes have long been synonymous with nomadic Scythians, Cossacks and Mongols galloping across ancient Silk Road trading routes, and the country's capital city seems incongruous in these desolate plains. Developed from scratch as a planned city in the early nineties, it has a prefabricated feel, as if it could be dismantled and carried off on horseback.

The new city was a personal project of Kazakhstan's strongman president at the time, Nursultan Nazarbayev. He poured large amounts of the country's significant oil revenues into establishing it. The government's most recent project has been to build the new Astana International Financial Centre (AIFC), to serve as a hub for global finance corporations and fintech start-ups. Nazarbayev now serves as AIFC's chairman, presiding over a governing board that includes financiers from J. P. Morgan, Citigroup and Russia's Sberbank. It is a perfect emblem of the intersection between state power and global finance.

I was invited to the Astana Expo 2017 – an international showcase for the city – by some associates who wished to use the occasion

Astana (Nur-Sultan)

to pitch themselves as advisors to AIFC. On the plane journey there I watched a Kazakh comedy film. It followed the hapless exploits of a broke hotel manager trying to get his rough, horse-riding relatives to leave their rural village and to come work as staff in his city hotel, where foreign businessmen stayed. Kazakhstan was one of the last great non-state realms, and the film set up a comedic showdown between the country's old free-flowing nomadism and its newer world of state oil companies and big finance.

The tension between settled statism and dynamic nomadism is an ancient one, seen the world over. For example, James C. Scott's *The Art of Not Being Governed: An Anarchist History of Upland Southeast Asia* argues that south-east Asia historically saw a battle between the two mentalities. The first was found in the valleys, where settled agriculture enabled accumulation and stable commerce to take place and hierarchical states with settled cities to form. The second came from harsher environments that states could not enclose, like mountains and deserts, where non-hierarchical semi-nomadic groups formed. The juxtaposition of the two allowed for *political exit* – if people did not like state-centred life, they could leave for the non-state realm, from where they could occasionally stage raids into the state regions.

Standing on the outskirts of Astana, one can imagine those days of political exit. The city abruptly finishes, and is replaced by seemingly endless plains to disappear into. It is this feeling that crypto promoters try to evoke when they talk about a large-scale exit from statism to the digital 'plains' offered by blockchain technology. One of the most infamous online dark market sites was called the Silk Road precisely to draw upon this imagery, as was its successor OpenBazaar.

Raiding can go both ways, though. In old times, states would launch expeditions into the inhospitable non-state regions to capture slaves. Alternatively, they could buy off the hordes, get them to raid each other, or hire them as mercenaries. Pirates could be pacified by giving them minor positions, and in recent years this 'pacification of the pirates' has become apparent in the crypto world. Entering the massive Astana exhibition centre for the Expo, I found myself watching a Texan blockchain consultant pitching the technology to Kazakh oil execs in the audience. He was from ConsenSys, a New York-based company that builds 'private blockchain' systems for banks and other corporations. Here were 'pirates', selling a privatised version of their technology to their apparent enemies. Let me take you through the gradual corporate takeover of crypto that led up to this, and how blockchain technology may strengthen corporate oligopolies, rather than weakening them.

Automating oligopolies

Early blockchain communities did have counter-cultural currents within them, but, as detailed in the previous chapter, many crypto

229

enthusiasts were also politically conservative and often pro-business. From the very beginning, crypto conferences felt as much like business networking events as political rallies, and by 2016 they had become highly commercialised, with big price tags for tickets and sponsors. By then trading crypto-tokens was a mainstream activity and, far from seeing it as a threat, investment banks like Goldman Sachs began flirting with the idea of starting crypto trading divisions to help their investor clients buy and sell the collectibles. Mainstream investment funds began to see crypto-tokens as one more thing to be placed into a portfolio, alongside stock market shares, bonds, real estate, rare art and gold.

This attention from mainstream players created ideological instability within crypto circles. In the early years, crypto entrepreneurs had presented themselves as enemies of big corporates, but – given that the profit motive is a political ideal in free-market thought – those same entrepreneurs later struggled to determine whether they were 'selling out' if those corporations offered them profitable collaboration opportunities. I got my first glimpse into this when I took part in a blockchain brainstorming session at a major UK bank. Several blockchain entrepreneurs were invited, and quickly sought to promote their initiatives to the bankers present.

That may seem counterintuitive. What 'collaboration opportunities' could even exist between banks and crypto rebels, and why would crypto entrepreneurs be pitching them blockchain technology if its purpose was to bypass the banking system? Isn't blockchain technology supposed to facilitate 'decentralisation'?

To understand what was going on – and what continues to go on – we need to remember that banks operate as oligopolies, and oligopolies are only semi-centralised, which means they are also semi-decentralised. The UK, for example, has five major commercial banks that dominate the market, with another five or so that fight for the remainder. Just as old royals would backstab each other while intermarrying to maintain a common front, banks half compete and half co-operate. They work together to maintain a common oligopolistic infrastructure, within which they then jostle for prominence. It is not dissimilar to Premier League football, which relies on much co-operation between the twenty teams that then compete within it. These systems are centralised from one perspective, but decentralised from another, in that they still involve a constellation of organisations that do not fully see eye-to-eye.

Much in the same way as Premier League football clubs have intense dialogue in order to arrange matches and facilitate deals (such as swapping players and so on), banks get together in the 'interbank markets' where they do behind-the-scenes wholesale deals with each other. The banks then have back-office teams to check that their records are the same as those of the counterparts they do business with, and to settle outstanding claims against each other. These interbank markets rely on a series of separate IT systems, run by separate banks, which are brought into sync by human teams.

What has this got to do with blockchain technology? In Chapter 11 I showed that blockchain systems are designed to *sync separate systems into unity*, so – with some alterations – a privatised version of the same technology could be used to sync these separate IT systems of banks into unity. Picture the Premier League co-opting a grassroots co-ordination technology used by an amateur league. Big corporates can raid and repurpose the technological scaffolding of blockchain technology to coordinate their cartels, syndicates and oligopolies.

This is what led to the explosion of interest in 'private blockchains', 'enterprise blockchains' and 'consortium blockchains'. Creating these systems entailed dropping various features of open blockchain systems that were unnecessary or undesirable in the corporate realm, and starting to call these watered-down systems by the more generic name of 'distributed ledger technology' (DLT). This became a buzzword in the mainstream fintech world. We've already seen how customer-facing staff are being automated away with apps, and how behind-the-scenes financial professionals are being replaced (or augmented) with AI. The new frontier involves using DLT to automate the activities of back-office interbank co-ordination staff, who cost financial institutions a lot without being overt profit-generators.

DLT is the next step in the banking sector drive for automation. This time, though, banks are interested in automating the co-operation processes between themselves. All sorts of financial insiders have piled in to build out 'blockchain' solutions for any form of interbank co-ordination, from securitisation to syndicated loans (when banks get together to collectively offer a loan). A collection of banks supported an initiative called R3 to build out this infrastructure, but many other collective projects have followed, such as Hyperledger. It is chaired by Blythe Masters, the former senior trader at J. P. Morgan infamous for pioneering the credit default swap market (which amplified the 2008 global financial crisis). A team from J. P. Morgan itself built its own DLT system called Quorum, and did so by raiding the Ethereum source code.

Not only do many crypto developers now moonlight as consultants for these projects, but it has also become commonplace for bankers to sponsor and be invited to blockchain events. Hard-line anarcho-capitalists (like those described in the last chapter) might grumble as Blythe Masters takes the stage, but – much like street cats may snarl at a passing tiger while secretly in awe – many respect the famous trader who made so much money for J. P. Morgan. Attention from such superstars is validation.

The media has sometimes reported on this coup as if it were a victory for crypto rebels, and as if banks were buying into the new decentralised world, rather than raiding it. The banks happily play along with this for free publicity, and this game now extends into almost all mega-corporations. All big corporates are part of the complex webs we considered in Chapter 1, operating as oligopolies with transnational supply chains that require co-ordination and co-operation. Microsoft's Azure cloud computing division now offers corporates the ability to run a 'consortium blockchain' among themselves entirely *within* the Microsoft datacentres. Microsoft is a member of the Enterprise Ethereum Alliance, set up to build a business-friendly version of the anarchic protocol, called Enterprise Ethereum. Our Texan who ended up in Kazakhstan pitching the technology to oil corporations belongs to the same alliance.

The explosion of profit opportunities has left the blockchain landscape complex and fragmented, and it is often not clear who is raiding whom: is the corporate world 'taking over' crypto, or is crypto taking over the corporate world? For members of the public the distinctions are becoming blurred. For example, the major crypto-token trading platform Coinbase is listed on the centralised NASDAQ stock exchange, while promoting trading in decentralised collectibles. At the same time NASDAQ has announced a partnership with R3 – the private corporate blockchain consortium – to help it build DLT infrastructure to service its corporate partners.

There are ambiguous zones of hybridisation occurring, but perhaps the most ambiguous is the emergent world of 'stablecoins'. These are becoming of crucial importance in debates about the future of our monetary system, and stand to disrupt that system in unexpected ways.

Decentralised promises for bank dollars

In Chapter 4 we saw how banks take our state money and issue us digital chips. Groups like PayPal, however, can subsequently take your bank chips, and issue you PayPal chips instead. 'Dollars' in your PayPal account are third-tier chips that promise you second-tier bank chips that promise you first-tier US government dollars issued by the Fed. PayPal chips get called 'dollars' because their issuer promises to redeem them for a 'higher-powered' dollar lower down the chain, thereby tethering them into the deep core of the dollar system.

There is, however, no inherent reason why a third-tier chip – similar to that issued by PayPal – cannot be pushed out on a blockchain network. This is exactly what a crypto company called Tether started doing in 2014. The company – like PayPal – would take bank dollars from people (which they put in Taiwanese bank accounts, backed by the US bank Wells Fargo), and issue them third-tier chips in the form of dark-market crypto-tokens. A Tether token was a 'crypto-dollar' backed by bank dollars, partially backed by the state dollars of the Fed.

These decentralised promises for bank dollars subsequently came to be known as stablecoins. The term 'coin' was supposed to evoke the idea of Bitcoin – thereby emphasising decentralisation – but with 'stable' added as a contrast. The reason a stablecoin is stable, however, is that it's just an extension to the normal monetary system. Most of what we call money is bank-issued or corporate-issued IOUs, and Tether was a crypto variation on this exact same principle.

Tether was founded by – among others – Brock Pierce, a charismatic serial entrepreneur (and formerly a child star in the nineties *Mighty Ducks* film franchise). Pierce made his original millions by trading in-game currencies on online games like World of Warcraft (and worked with Steve Bannon on these gamer currencies), but later shifted his persona into a hippy libertarian crypto-guru. In 2017 he set out to develop 'Puertopia', a 'crypto-utopia' in Puerto Rico, painting a glamorous picture of the island as a future low-tax playground for jet-setter crypto-entrepreneurs.

Tether, however, became embroiled in controversy, because its bosses ran another crypto company that lost millions in a major hack, after which they allegedly raided Tether's bank account reserves to cover those losses. This led to Tether tokens being partially 'unbacked' – or *untethered*, if you will – for some time. The controversy led to Wells Fargo pressuring the Taiwanese banks in 2017 to ditch the Tether company, so Pierce set up a bank in Puerto Rico to serve as a back-up to which Tether's US dollar reserves could be moved. The story has been through many iterations since then, but Tether's crypto-dollars circulate freely while the bank dollars that back them have been moved to different banks in different locations. At the time of writing there are some $69 billion worth of Tether tokens circulating in the world, which means there is (supposedly) $69 billion in the company's bank accounts, wherever those may currently reside.

In their early iterations, stablecoins could be characterised as a raid by crypto-pirates into the world of fiat money, but by 2018 many other companies started getting into this business and taking it mainstream. The major crypto company Circle, for example, also uses the business model inherited from PayPal, taking your bank money (and the interest you might otherwise have earned on that), and then issuing a crypto-dollar called USD Coin to you in return, via an Ethereum smart contract. Other stablecoins such as DAI have a more advanced mechanism for associating themselves with the US dollar, using *pegging* to make their token mimic a dollar without being directly tied to banks. These stablecoins are interesting in the context of the war on cash: unlike Bitcoin, they are usable as a stable form of money, and – given that they can be used (semi)-pseudonymously – have a better claim to being called 'digital cash'.

But the stablecoin concept itself can be re-raided. Earlier I showed how the corporate incursion into blockchain technology resulted in private DLT systems, and these can now be used to implement a carefully controlled semi-centralised corporate stablecoin, not dissimilar to a centralised PayPal. This is precisely what happened in 2019, in one of the biggest money stories of the year, and one that still looms over us.

Big Tech raids the stablecoins

A dollar-promise can be rebranded. A casino chip that says 'Caesar's Palace Casino' is a rebranded US dollar promise, as are countless privately branded 'virtual currencies' from various corporations. For example, Amazon will allow you to transfer US dollars into its bank account and in return issue you Amazon Coin, which can be used on its platform like a digital dollar voucher. A more advanced trick is to issue such a voucher in response to receiving *multiple* underlying currencies. Imagine, for example, Amazon receiving transfers from ten different people, in ten different currencies, into ten different bank accounts belonging to Amazon in ten different countries, but then issuing them all the same digital voucher for their proportion of the bundle. That voucher is backed by ten different currencies but, rather than trying to brand it as such, Amazon could pretend that it is free-standing by giving it a brand name like GlobalDominationCoin.

Amazon has not tried to do this, but Facebook has. In 2019 it launched a raid on a project that I was involved in called DECODE, an EU-funded initiative to build citizen-owned digital platforms as an alternative to Big Tech. So much for that. Facebook lured one of our core technical teams away and hired them to design something they branded as Libra. If the marketing was anything to go by, it was going to be a global 'cryptocurrency' to save the world.

Libra was the culmination of all the trends covered in this book so far. Facebook explicitly joined the war on cash, demonising physical state money as an outdated relic holding people back, and attached that to a message of financial inclusion, claiming that its new system would help the world's unbanked. Like the fintech sector, it romanticised automation, but simultaneously sought to leech off the edgy pirate aesthetics of the crypto movement by claiming that Libra would be a 'cryptocurrency'. In reality it was to be a private DLT system, controlled by a syndicate of corporations, operating through a Swiss non-profit association that had various bank accounts to hold various currencies. People across the world would be able to transfer currency into those in exchange for a new Libra unit recorded on a private consortium blockchain controlled by the corporate members. It would be a corporate 'stablecoin', backed by a range of global bank chips.

The corporate members included Silicon Valley super-powers like Uber, and the role of these partners would be to agree to accept the new Libra token, which would supercharge the token by giving users a reason to hold it. Think of these tokens as vouchers redeemable with a range of mega-corporations. Uber itself would not *use* Libra – it needs actual US dollars to pay its shareholders – but would take any Libra it received from ordinary people back to the association to redeem it for bank-money, after which the Libra unit would be taken out of circulation, or 'burned'.

Co-ordinating this effort was David Marcus, a former president at PayPal who had since moved to run Facebook's Messenger division. While we think of Facebook Messenger as a place to informally message friends, digital payments also entail sending messages. A normal Facebook message might be, 'You want to grab dinner?', addressed to Sanjev, but a sub-section of the same messaging system could be calibrated to send specialised financial messages like, 'Transfer 150 Libra units to Sanjev'. Facebook figured that – even if it did not unilaterally control Libra – it controlled some of the world's biggest messaging systems, like WhatsApp and Messenger. If people opened WhatsApp and saw a tab within the app that said 'Transfer Libra', hundreds of millions of people could be steered into the system.

To further distance itself from suspicion of its commercial motives, Facebook enlisted NGOs and microfinance groups like Kiva to be inaugural members, thereby presenting Libra as a humanitarian life-raft for the world's 'unbanked'. Libra's spokesman Dante Desparte spoke in eloquent tones about his personal dream of empowering millions through financial inclusion: if the whole world could be on-boarded into one huge system – rather than relying on smaller systems tied together – the dreaded menace of *friction* would be defeated once and for all.

A wave of media reporting took it much further, with headlines claiming that Libra would 'destroy commercial and central banks' or become a 'global central bank'. This latter notion was the inverse of reality. A central bank is an institution that issues money that commercial banks use as reserves against which they issue their own bank chips. Libra, though, was *backed* by bank chips. It was to be a third-tier chip system built on top of banks, not a central bank behind banks. For it to be a central bank, it would have to be the monetary centre of gravity, with banks using its Libra units as reserves to back themselves. This would require the entire global monetary system to invert, which is not about to happen.

Nevertheless, if Facebook was successful in rebranding a promise for all the world's currencies in its own colours, it could conceivably become a monetary power of sorts, at least at the psychological level. But while it is true that the Libra unit was to be backed by national currencies, it was uncertain as to which of those currencies would be used, and in what proportion. For example, if large numbers of Indian users transferred rupees to the Indian bank account of the Libra association, the association might exchange those for US dollars. The Indian user would see a Libra unit, while in the background their national currency got traded on foreign exchange markets by a Swiss foundation.

For all those post-colonial societies already struggling to maintain a sense of monetary sovereignty in a world dominated by the US dollar, yuan and euro, this was an unnerving prospect. Libra would certainly be backed by standard cloudmoney, but probably that issued by the banks of the monetary superpowers. The Libra narrative painted this as a good thing. Not only was Libra supposed to save the world's unbanked from cash, but it was also supposed to save them from their dodgy national currencies. It was presented as a broad, stable meta-currency to combat individual hyper-inflating currencies, such as those historically found in Zimbabwe. This is where for me the story got even more personal.

Zuckerberg vs Mugabe

My parents are Zimbabwean, and I spent much time in the country growing up. When my parents were children the country was a British colony called Rhodesia. In 1965, though, white Rhodesians broke away from the Empire and formed a white-run settler nation. In doing so they rebranded the pound there, in an act of defiance, as the Rhodesian dollar. During this same period the government imprisoned a schoolteacher named Robert Mugabe for anti-government agitation. After a decade of imprisonment he was released, and set about building a black rebel movement, which ended up pitted against people like my father in a brutal bush war. When the rebels won power in 1980, Mugabe rebranded the Rhodesian dollar as the Zimbabwe dollar, which imploded in the 2000s as Mugabe lapsed into dictatorial paranoia.

Zimbabwe's failed currency is now frequently held up as a

cautionary tale about the fate of fiat currencies, but such tales ignore the geopolitical inequalities between states. Post-colonial Zimbabwe was a country relying on commodity exports, with a large population of poor black farmworkers alongside a small population of wealthy white farmers. It is not surprising that internal tensions emerged from this, and that the situation was exploited by opportunistic leaders, resulting in the country's agricultural sector collapsing amid an unstable political environment. In this context, a weak government tried to magic away the situation by issuing more money, but – using the nervous system metaphor we established in Chapter 1 – that is like sending activation impulses to a lame arm. If a country's underlying productive system is collapsing, that needs to be stabilised before any type of monetary interventions will work.

In this situation Zimbabwe's currency ended up unravelling, so its 2009 emergency plan was to abandon its own currency network, and leap to US dollars. The network relationships that hold the US dollar together are extremely strong, with a domestic US population of more than 300 million people that command high technology and huge geopolitical power, secured by massive military and cultural influence.

Mugabe might nowadays be reviled in the West as an authoritarian dictator, but nobody can claim that the currency system he drove into the ground was not rooted in the place he came from. For a post-colonial nation, allowing its own currency to collapse and being forced to use an imperialistic one can be humiliating. Zimbabweans do not desire to be subjects of the US Federal Reserve, and feel no affinity with the old men printed on its dollar bills. The same can be said for the resentment felt towards the US in dollarised countries like Ecuador. The primary concern for democracy activists in such countries is not whether money can be transferred using a slick app. It is whether their governments are able to carve out some degree of political self-determination, rather than being client states of great powers.

The way Facebook showcased Libra showed either ignorance of, or lack of interest in, such aspirations. It presented a vision of a monetary system which would offer convenience to the likes of poor Zimbabweans, but without any anchoring in their political context. Centrist thinkers were easily seduced by this presentation, because the Facebook oligarchs are smooth-talking technocrats rather than old-school autocrats like Mugabe. Behind Libra, though, was a corporate consortium with no democratic credentials of its own, likely to back its unit with powerful currencies like the dollar. It was never intended to help Zimbabwean democracy activists rebuild their own institutions: the point was to get them to rely on a transnational corporate infrastructure.

It turns out that Libra's seductive simplicity did generate suspicion. What hidden movements between banks might the Libra association trigger when a Libra token moved 'seamlessly' across borders? The opacity caused backlash from the world's regulators. Members of the US Congress sent a letter to Facebook asking it to officially halt development, and David Marcus was called in to testify. He quickly reverted to a geopolitical justification, arguing that Libra would be a US counterforce to Chinese digital payments platforms (a message that Facebook CEO Mark Zuckerberg pushed too). Libra could be like a financial weapon, ensuring dollar hegemony extends beyond the realm of international trade, and into ordinary retail transactions between citizens within foreign countries. Nevertheless, the backlash towards its platform caused Facebook to back down and return to the drawing board. In 2021 it rebranded Libra as Diem – planning to issue standard US-dollar backed corporate stablecoins like Tether – but the US government blocked Diem from being developed further, forcing Facebook to sell the technology to a new stablecoin contender called Silvergate.

The state raids everyone

We have seen commercial banks raid the blockchain concept to co-ordinate their oligopolies, and we have seen crypto-entrepreneurs raid the tethering concept to create so-called stablecoins. We have then seen corporations like Facebook raiding the results of both to create oligopoly-run stablecoins. Now it is time for central banks to stage their own raids.

Many digital money systems have tried to draw upon the emotive appeal of physical cash by calling themselves 'digital cash', but all have failed to replicate cash on one or more dimensions. Bitcoin has left us with a movable pseudonymous token, but it largely fails to operate as money, while stablecoins get closer but remain third-tier chips dependent on banks.

What we call cash is anonymous physical legal tender issued by states, which suggests that state institutions are in the best position to issue a digital version of it. In Chapter 3 I noted that there already is a form of digital state money available to commercial banks, who gather to use it alongside state institutions (like the treasury) at the central bank. This digital version of state money (called 'reserves') is like the internal currency of a private membership club, while the rest of us use the public physical version.

But what if the central bank opened up that club? Imagine if you could walk up to the Bank of England in London, the Fed in New York, the ECB in Frankfurt or the People's Bank of China in Beijing, and open an account with them. Imagine if, after opening that account, they gave you a great Internet banking platform – just like a commercial bank would – and a payments app. You may not need to imagine this much longer, because these central banks are all considering the idea.

While the private club version of state digital money gets called reserves, this proposed public form gets called 'central bank digital currency', or CBDC. It is the war on cash that is partly responsible for inspiring this idea: in Sweden, for example, the Swedish authorities are worried about the resilience of their hyper-digitalised money system, so have begun investigating whether to issue a CBDC called the *e-Krona*. They realise that if cash were to disappear, a lot of people would be at risk because the commercial banking sector – with its profit-motive guiding its actions – cannot be trusted to serve less profitable members of society.

Furthermore, they are aware that the decline of the physical cash system implies the rise of the digital bank system, alongside media-grabbing initiatives like Libra and mega-platforms like Amazon that rely upon that digital money infrastructure. Central banks are getting edgy about the growing power of these players and feel pressure to appear to be 'keeping up' by matching the actions of the fintech world. Over the last few years the CBDC concept has percolated through the central banking community, and in several countries teams have been set up to investigate if such a system is feasible or desirable.

What would be the difference between a CBDC and, say, PayPal, or Venmo, or any other unit moved around using a fintech payments app? With Venmo the unit you are moving is a third-tier corporate chip, but with a Bank of England app you would be moving a first-tier state one, much like you do when handing over cash. Venmo can go bankrupt, rendering its promises useless, whereas central banks generally do not suffer from this problem.

But this could cause other problems. The prospect of holding risk-free central bank money might lead to tens of millions of people queueing up to open accounts with a central bank. Not only is this an administrative nightmare, but it might also induce people into fleeing from commercial banks, reducing those banks' profitability through a 'digital bank run'. A normal bank run occurs when people panic and rapidly convert their bank chips into state cash at an ATM or branch. A digital run would entail people instructing their existing commercial banks to transfer digital state money into their newly opened central bank accounts, draining the reserves of those banks.

One major central bank think-tank surveyed twenty-three central banks, nineteen of which expressed concern about CBDCs causing digital bank runs. In the same survey, some poorer countries suggested a need for a CBDC to compete with private sector US-dollar-heavy challengers like Libra, which are already threatening to attack their monetary sovereignty and domestic banking systems. If you are a central bank governor in a small country where your citizens use apps controlled by US corporations, you face the prospect of those apps being used to fob off US dollar corporate stablecoins on them. Maybe the only means to compete is to offer your citizens direct access to your central bank as an alternative, even if it negatively impacts your domestic banking sector.

Traditional cash is tactile and anonymous. CBDC, by contrast, would suffer from all the same issues that surround digital money more generally. This includes potential surveillance, censorship and the continued enabling of creeping corporate capitalism (Amazon can still do business with a CBDC better than it can do with cash). Some of these issues could be addressed, though, if a central bank were to scavenge parts from the crypto world, implementing a CBDC using a private blockchain system in which people hold state money in pseudonymous addresses. This would come closer to true 'digital cash'. A state digital currency in this vein might even serve to break the power of the corporate banking sector in places where it dominates.

These possibilities lead to a new contradictory set of problems. Traditionally there is a power-sharing agreement between central banks and commercial banks: the former issue resilient, inclusive and offline public cash, while the latter issue private digital money that spies on people. This balance can work, because while a small percentage of cash can end up in criminal markets, it is slow and cumbersome to use within transnational black markets. If a central bank's hand was forced into issuing an anonymous CDBC, international cybercriminals would suddenly have access to a form of transnational state money that is much faster than cash. The non-anonymous alternative, though, is a surveillance nightmare: this is what we currently see being proposed in China, which is going full speed ahead with trials on a new CBDC that could be integrated into its social credit system. Imagine streets lined with facial recognition cameras that can identify people breaking a curfew during a pandemic, and which send a message to automatically deduct a fine from a person's account. Technically feasible? Yes. Politically possible? Increasingly. Desirable? Well, that really depends on who you are.

This also has cross-border ramifications. Imagine if Zimbabweans could download an official yuan CBDC wallet app from the Google Play store and use digital yuan within Zimbabwe. This is the fear that Facebook was trying to tap into when arguing its Libra case with US lawmakers. It is also why USAID partners with groups like Visa in its efforts to digitise the Indian monetary system. The US and China are both geopolitical and corporate powerhouses, but the US sometimes prefers to carry out its foreign policy via private corporations. Perhaps the US government may come to be swayed by the idea that corporate-issued US dollar stablecoins should be promoted to my Zimbabwean family members, lest they begin using the digital yuan.

Where does this leave us?

Blockchain technology certainly has disrupted the public narrative around monetary systems, and introduced many new terms into the debate – but has it led us anywhere fundamentally new? Bitcoin tokens remain largely perceived as a collectible investment to be bought and sold like any other, while the underlying technology has been co-opted into co-ordinating oligopolistic cartels, which is hardly novel in the grand scheme of capitalist history. Stablecoins and CBDC certainly do cause waves in the monetary system but, despite their differences, all these digital options have one feature in common: they can be integrated into the operations of Big Tech, and can even be issued and directed via Big Tech platforms.

The most contrarian form of money in the world, then, may be the simplest and most unfashionable. Cash. It remains a bug in the system, standing in the way of the fusion between finance and tech, and this is precisely why it is one of our last hopes.

Conclusion

In South Africa I experienced years of 'rolling blackouts' when the electricity infrastructure failed. Electricity animates so much of modern life, so as the power went out towns were plunged into darkness, televisions involuntarily cut out and all appliances became unusable. This caused much disruption, but it also had another effect. With the hum of electricity gone, we would hear the insects, and the blanket of darkness allowed us to see the stars. The Internet would go too, cyberspace giving way to real space.

Each day we may be distracted by the blitz of 24-hour news and social media, but when the power goes out we're forced to think about the foundations of our existence. A hundred thousand years before the arrival of market systems we lived under the stars in tight groups, relying upon entirely different principles to survive. It is only in the last few thousand years that monetary systems have become embedded in our communities. And it is only in the last few decades that those have grown into an interconnected digital mass, reaching out from London and San Francisco to the Amazon rainforest and rural Kazakhstan. The next decade may see this thicken into a dense, inescapable mesh.

I began this book by rejecting the popular metaphor of money as blood circulating around an economy. Instead I presented monetary systems as being akin to a nervous system embedded in communities of people. And, while this enables all of us to activate each other into motion, the power concentrates within large players. It approximates at a global scale to the situation found in our own nervous systems, which are split into a central nervous system and peripheral one. The central nervous system is the domain of the brain, which has a motor cortex that translates our thoughts into action, sending out impulses that make our limbs move. If you imagine the global economy as an interconnected body of workers and natural resources, financial centres (like London, New York, Singapore, Shanghai, Tokyo, Frankfurt, Dubai, etc.) form a transnational 'motor cortex'. Collectively they have the capacity to induce mass action.

In the lead-up to the financial crisis of 2008 it became apparent just how misguided this 'motor cortex' can be. Hundreds of thousands of workers were pushed into building houses that would stand empty. Many more impoverished people were pushed into debt to buy them, and their hopeless promises to pay back bundled into packages sold to mega-funds across the world. The financial sector created a situation in which the global economy took on the appearance of a veering drunk losing their motor functions. But even when the sector seems to operate normally, the general direction of travel in our global economy is like a dazed drift towards a cliff edge. We are using up our planet's resources in a large-scale yet short-term drive for expansion and profit, while power centralises and inequality deepens. All major corporations are trying to accentuate and accelerate that situation, rather than slow it down.

But while the central nervous system of money clusters within major institutions, on its peripheries – in the form of cash – it gains some distance from them. Cash forms a peripheral nervous system, partially disconnected from the central one. As cash 'impulses' exit an ATM, they must conduct themselves through human beings in geographic space. In the dark of night, when the electricity fails, people in Amazonian Pucallpa can still trigger each other into action by passing small coins.

Such peer-to-peer human 'conductivity', however, hampers the drive towards acceleration that lies at the core of our economic system. Corporate capitalism has a constant inbuilt tendency to expand, and to do so must pull the world into a transnational mesh. This tendency now requires it to destroy the peripheral system of cash, in favour of money contained within a centralised conglomeration of data centres. Those reach out to us via digital finance infrastructures, and increasingly tie us in via the collaboration of technology corporations. Big Tech and Big Finance are becoming like two hemispheres of a global mind, fusing through their natural synergies.

As I write this in mid-2021, this fusion is more apparent than ever. A global pandemic has raged for a year now, and the fintech sector has weaponised it to demonise cash even further, presenting it as slow and dirty. This has led to large parts of the retail sector urging shoppers to drop cash in favour of digital payments. In the UK this has worked, with cash usage plummeting by over 50 per cent in 2020. The automated finance industry has hit a bonanza, and has done so alongside its natural allies: in 2020 Amazon's revenues rose by over 40 per cent as it raked in almost $14,000 in digital money *per second*, a feat closely followed by all the other big tech firms.

It is physically impossible for a human to do any form of labour that could possibly earn $14,000 per second, but these systems can do it because they are not human. They are corporate infrastructures that have seeped between our interactions, from where they can harvest inhuman levels of revenue to steer towards small groups of shareholders. Thus, unlike children in 1921, children born in 2021 will first encounter the concept of 'Amazon' as a colossal digital logistics empire for mass consumer goods (operating in conjunction with digital finance), rather than a rainforest that sustains human life.

My primary goal in this book has been to show that Big Tech and Big Finance now collaborate to steer us into an overall system of corporate capitalism that remains as toxic as ever, albeit now faster and more convenient to interact with. My secondary goal has been to highlight the dangers of this situation, including the potential for mass surveillance, censorship and manipulation. My third goal, however, has been to convey the contradictions of this situation, and our conflicted feelings about it. This is both a broader and more nuanced area of inquiry than the previous two, and those contradictions will by nature remain unresolved. In the final stretches of this book I would like to address them more closely, after which I will discuss paths we may take going forward.

The contradictory bind

When describing the rise of automated surveillance capitalism, it is easy to point out its various dangers, but something more subtle drives my own discomfort. It is the pervasive feeling of inauthenticity that accompanies it. It is that tremor of emotional conflict a person feels when – in full knowledge of how Amazon is taking over the world – they nevertheless sense the futility of resistance, and find themselves with their finger on the 'Buy' button.

Much of our public discourse fails to address this inner conflict. Rather, we get two stories, one conveying agency and another coercion. Many mainstream futurists and economists veer on the side of agency, portraying the rise of digital giants as driven by the desires of empowered consumers who love convenience. The technology critic, or conspiracy theorist, by contrast, paints an ominous picture of us being coercively herded onto platforms. While I veer towards the latter neither story fully satisfies me. My true interest is to uncover the contradictory *simultaneity* of agency and coercion. I focus on the dynamics of a 'cashless society' because it is a cutting-edge example of how individuals experience themselves 'choosing' to use something – digital payments – that simultaneously appears as a pre-destined inevitability chosen for them.

I believe that the key to understanding this tension is to see that, within large-scale interdependent markets, agency and coercion are inseparable. Since the eighteenth century, economists have hinted at this when speaking of the 'wisdom of the market'. The phrase conjures imagery of a meta-entity generated by people and yet transcending them. This is where Adam Smith's concept of the 'invisible hand' emerges: the actions of people within an exchange network push and pull upon the other members of that network, via the monetary system that holds it together. This is not a vision of individual freedom. It is a vision of collective entanglement, albeit one that people experience as individuals.

Economists like Smith were writing in a time when nation states were disbanding small pre-capitalist communities and transforming them into large networks of strangers held under common money systems. Seen from one perspective, money was separating people – people in large markets do not know each other – but seen from another it was forging an abstract connection between those who might not otherwise interact. This unlocks a peculiar mindset: as people become dependent on distant others they cannot see, they become disassociated from the origins of the things they use, and the destinations of the things they produce. In pre-capitalist society, individuals were aware of their kin, and the ecologies they depended upon, but the further we travel from that awareness the closer we get to a mindset of decontextualised commodification, where value becomes conflated with the money used to co-ordinate a market.

In our nervous system metaphor, we may imagine monetary systems spreading out to connect distant people, while older, pre-capitalist forces – like those cultures of reciprocity and ecological awareness – keep the toxic sides of markets in check (almost like pain receptors). But the vast market systems that monetary systems unlock can also detach from the communities and ecologies that underpin them, to become numb and invasive. Indeed, the core feature of modern large-scale capitalism is that frenetic acts of monetary exchange become exalted as spirits in and of themselves, which must accompany – and dominate – every other part of our being. This soul-less drive to maximise transactions leaves a growing existential void.

We have blindly stumbled into systems that exploit our short-term desires to the detriment of our longer-term ones, and they *break* and disrupt our lives if we attempt to pull back from them. Rather than crawling towards Utopia, then, large-scale markets crawl towards concentrating production and consumption into pure conglomerations of profit-seeking, represented most acutely by transnational corporations. While the individuals who work within this conglomeration can *feel* many things, the financial and corporate sector as an institutional complex is unable to 'feel' any-thing except profit, so our systems are running away with us, like a centrifuge spinning ever faster.

It is this which leads to visions of a giant technological 'Singular-ity'. Google's futurist-in-chief Ray Kurzweil tries to put a mystical spin on this by invoking the Enlightenment tradition, which sees history as one long march to human transcendence over nature, alongside a parallel ascendency of mind over body. It begins with a vision of us crouching naked in the prehistoric wilderness. It ends with us ascending into a human-technology hybrid that lives for-ever, colonises space and regulates an artificial environment at will through AI 'super-intelligence', a pure, transcendent, rational mind. Displayed prominently above the washroom urinals at Sin-gularity University is a comic strip based on a sermon given by Kurzweil, with quotes as follows.

252

Evolution is a spiritual process. Technological evolution is the same as biological evolution. In the future we will becomes a mix of bio-logical and non-biological intelligence . . . plugging our brains into the cloud, effectively expanding our neocortex. Becoming closer to God, the Ideal. Humans plus AI as one.

Let's face it. The transcendent spiritual AI cloud Kurzweil and his associates are referring to is the financial-corporate conglomeration I have been describing. The 'state-of-mind' of this conglomeration is that which I described in Chapter 1 – the cold logic of skyscraper-bound legal entities scanning through people for profit opportunities at scale. It is an outlook calibrated to over-value large-scale friction-less efficiency to the detriment of the deeper things we love – unknown wild spaces, peer-to-peer connection, texture, spontaneity and unguided journeys. If corporate capitalism was allowed to fully express itself, it would demand access to our very brainwaves, pro-moting payment-by-telepathy to access the thoughts of others. This creep of hyperconnected markets into the deepest parts of being is the defining feature of our age.

Perturbations within business-as-usual

I began this book by promising to ignore 'wave-riding' stories and to explore the plate tectonics of the global economy instead. This creep of hyperconnected markets – mythologised by the likes of Kurzweil – and the fusion of Big Finance and Big Tech that is facili-tating it, is the underlying force defining the future. And yet it is barely mentioned on a day-to-day basis, because it is taken for granted, and public debate is focused instead on the perturbations found within that overall trend. Certainly, a maelstrom of commer-cial and geopolitical struggles continues to play out as the different mega-players jostle for relative power over the international payments infrastructure. In the process, Western and Eastern spheres of influence are emerging, with a US tech-finance conglomeration defining itself in opposition to a Chinese one (while the European Union struggles to form a third force midway between them).

These ongoing tensions provide interesting possibilities for unexpected changes in due course. For example, while digital finance might have had a major win during Covid, the commercial banks that provide the underlying digital money find themselves in a delicate position. Their pandemic success may have been too rapid, making central banks nervous. Central bankers know that if they let cash die, it will result in commercial bank domination, but if they push CBDC as a replacement they could destabilise those very institutions. Central banks are not in the business of putting commercial banks out of business. If, however, they do not provide a form of 'digital cash', they risk creating public demand for dark-market alternatives like the stablecoins Tether provides.

Over the coming years we are likely to see the mega-players getting locked into a Mexican stand-off (the phrase refers to a situation in which no party can initiate aggression against another without triggering some threatening situation for themselves). The financial giants have undermined cash, but that now risks inducing central banks to compete with them digitally, or alternatively to allow stablecoin players to rise. One way to break this stand-off would be for central banks to step up and promote physical cash, but that would require going up against the much more powerful force of transnational corporate capitalism, which transcends individual states. Remember, for example, how Amazon has attempted to lobby against pro-cash legislation.

This all bleeds into the geopolitical realm, where nation states attempt to forge advantages for their own transnational corporate actors. The Chinese state has a clear outward-focused mission to expand its tech giants, and is simultaneously making moves to push ahead with a CBDC. Unlike the US, which prefers to maintain a tenuous distinction between its corporations and the state, CBDC poses less of an ideological conflict for China – the state already has strong control over its banking sector. But, given that the global economy is transnational, the Chinese CBDC is a threat to US interests. The US remains ideologically bound to the idea that it represents free markets, so it is likely to promote itself through players like Visa and Facebook, via which it can make inroads through WhatsApp into countries such as India. These transnational companies, however, are channelling a spirit that transcends the nation states they base themselves in, and they are opportunistic: if Facebook feels the need to let Indian WhatsApp users, for example, pay with a stablecoin backed by the Chinese CBDC, who is to say it will not do so?

These messy commercial geopolitics will continue to unfold over the next decade, but within the boundaries of the overarching Tech-Finance fusion. Most citizens do not feel any personal agency when it comes to defining how this fusion will play out, but the intuitive desire to fend off the creep of hyperconnected surveillance capitalism still drives many counter-reactions. In this vein, we should expect to see our cryptocurrency 'protestants' continue to imagine an exit to a non-state realm in cyberspace. This vision is often misguided and conflicted, but has some silver linings, so let's now turn to that.

The strengths and weaknesses of crypto cargo-cultism

In the early twentieth century, anthropologists observed a strange phenomenon in Melanesia. Islanders were building wooden aeroplane replicas in the hope that they would lead to the appearance of cargo payloads floating down from the sky. The anthropologists called these *cargo cults*.

Some scholars have interpreted these groups as a response to the stress of the colonial era. Foreign aeroplanes were associated with freight deliveries, so to islanders with little experience of the chains of causation it might have felt logical to build makeshift imitations. Cargo-cultism involves producing out-of-context artefacts (a wooden aeroplane) in an attempt to bring forth a world associated with them (abundant cargo): it is not irrational *per se*. It is just the wrong sequencing of events, with cause–correlation confusions.

Cargo-cultism – the act of building the surface appearance of something without the substance – is rife in many idealistic communities, including the Bitcoin movement. That movement continues to claim that limited-edition cyber-collectibles will challenge the global monetary system, but it is plainly apparent that crypto-tokens are being swallowed up by standard markets, and transformed into products to be traded like anything else. Indeed, the only reason why crypto-tokens seem 'money-like' is that their dollar price – forged in these speculative markets – enables them to be *swapped* (countertraded) for other things with prices. If you remove the dollar price, this ability ceases to exist. Put differently, it is their dollar price that enables them to have money-like qualities.

Hard-line Bitcoiners see this assessment of Bitcoin as a critique, and yet Bitcoin's swap-ability may usher in *a new age of countertrade*. Countertrade – the act of swapping money-priced goods with each other, rather than using money itself – has historically been limited, but Bitcoin (and other similar crypto-tokens) can supercharge this practice. This may indeed prove to be useful for people living under difficult regimes, or for those operating in the cracks of mainstream monetary systems. The weakness, however, of relying upon such countertradeable collectibles is that a single tweet by billionaire industrialists like Elon Musk can randomly and violently alter their price. This is extremely disruptive to the lives of any vulnerable people attempting to use them.

There also continues to be a conflict between the reality of Bitcoin and the marketing rhetoric put out by the growing industry that surrounds it. The latter markets the crypto-tokens as a competing monetary system, while simultaneously fixating upon its dollar price. To resolve this cognitive dissonance, industry cheerleaders claim that the rising price will culminate in a future inflection point where the entire monetary system will invert, leaving everything priced in Bitcoin. This is a category error. Nobody believes that the rising price of land, Amazon shares, rare postage stamps or even gold will lead to those being used to price things. No, a rising price is far more likely to make someone hold onto something in anticipation of future monetary gains, than it is to make them perceive the thing as money itself. Nevertheless, this practice of marketing crypto-tokens as 'money' is now commonplace, and it is leading many people astray by giving them false hopes.

Perhaps, however, people *need* false hopes. One strand of anthropology that may offer an insight into understanding this phenomenon is the study of carnivals. Anthropologists have studied how carnivals – ancient and modern – allow temporary upheavals of the social order in which people can let off steam by engaging in a temporary fantasy of escape. And Bitcoin token speculation appears very carnivalesque. The existential emptiness that accompanies corporate capitalism can generate both rebellious feelings and a yearning to get rich quick in order to escape it. Bitcoin enables people to act on both simultaneously: they can imagine themselves overturning the economic order while pursuing speculative dollar gains in that very same system.

The Bitcoin movement is not the only community of troubled technological dream-weavers. The Ethereum community, and the broader blockchain movement, continues to forge ahead with its smart contract systems and automated DAOs. We've seen how these are easily co-opted by corporate oligopolies who raid the technology, so that what is referred to as 'blockchain technology' now contains a confusing amalgamation of disparate agendas. Still, interesting hybrids continue to emerge from that meeting. Like fintech promoters, many blockchain innovators are often pro-digital automation and openly anti-cash, but they claim a desire to build a decentralised version of the finance-tech fusion that is sweeping the world. If cloudmoney is digital money in mega-data centres, then blockchain innovators imagine themselves flattening the cloud to create a digital money infrastructure administered in a more distributed form.

That vision is partially fantasy, because the greatest hope for the Ethereum (and other crypto) communities lies in stablecoins, which remain tethered to, or pegged to, the normal monetary system. Stablecoins are now being used to build so-called 'DeFi', or 'decentralised finance', platforms. Much in the way ordinary fintech companies rely upon digital bank money, DeFi involves setting up smart contract systems that will administer, lend or route digital stablecoins, and thereby replicate – in a more decentralised form – the same processes of financial automation that mainstream fintech promotes. A DeFi platform, for example, may seek to push out stablecoin loans, providing a partially decentralised fintech platform.

In blockchain technology circles this term 'decentralised' often means 'a large, distributed infrastructure controlled by nobody in particular'. In broad terms, then, the blockchain technology movement proceeds under the assumption that building transnational digital infrastructures that are usable by everyone, but controllable by nobody, is a worthy goal. But this conflicts with a more ancient tradition, in which 'decentralisation' is understood to be the process by which large infrastructures are broken up in favour of smaller ones that can be managed locally. This is the tradition behind many

258

modern alternative economy projects that attempt to *re-localise* economic life, by, for example, building small-scale permaculture projects, community credit unions, local co-operatives and community currencies. Behind these lies a vision of a world in which a constellation of community initiatives is federated together in harmony. It's a different ethos to the one that underpins crypto-tokens, which are imagined to facilitate a faceless transnational dark market where nobody needs to trust either institutions or each other.

Is there a way to harmonise these disparate visions of decentralisation, and to combine the strong elements of each into something more powerful than either can offer alone? The answer is yes, and it involves the use of promises.

Promises are the most primal predecessor of money: consider how your friend is prepared to give you something now under the implicit promise that you will return it – or something like it – later. As you draw upon a friend's hospitality, they develop an implicit positive 'balance', in the sense that you will feel a need or desire to *reciprocate*. When you draw upon favours, goods or support from friends, they – in a very informal and unmeasured sense – find themselves with positive 'credits', while you will find yourself seeking to bring the balances back to equality, or 'zero', as it were. These systems of reciprocity hold communities together, but in this simple interpersonal example we see the keys to 'people-powered' monetary systems: the promises close friends give each other are informal, unrecorded and unmeasured, but it is not hard to imagine that a more distant associate could hand you something in exchange for a more formal, recorded promise. The most well-established tradition of this 'paying by promise' is called *mutual credit*. Mutual credit systems – such as the Sardex system in Sardinia – are networks of people who issue promises to each other to obtain goods and services, while a central administrator sets standards, resolves disputes and keeps track of the promises each person has made. Each person's balance of promises expands or contracts, depending on how much they give to others, and how much they draw upon in return.

Mutual credit systems are a little like mini-nervous systems, often connecting mere hundreds of people together, rather than hundreds of millions. Such 'pay by promise' systems, however, can be scaled up by repurposing crypto technology: rather than using crypto networks to pass around abstract cyber-collectibles like Bitcoin, they could be used to pass around organic interpersonal promises. The 'six degrees of separation' theory suggests that you can reach anyone in the world by hopping through six 'friend-of-a-friend' connections. If I know you, and you know someone else, I can issue a promise via you to that person, in order to obtain real things from them in return. This concept is called *rippling credit*, because promises can ripple between people who trust each other. Of all the concepts that could create an entirely parallel monetary system, this is perhaps the most profound. As this book goes to print these, and related, ideas are being pushed forward by initiatives such as Trustlines, Circles, Sikoba and Grassroots Economics, who are implementing these concepts using blockchain architectures. For any person looking to get involved with idealistic-yet-practical currency innovation, this zone of hybridisation is the place to be.

Meditating on, and with, cash

I don't, however, want to leave readers with a false sense that all we need do is build community currency systems, and it is wishful thinking to expect that alone they would have the power to pull against the tech-finance vortex engulfing us. This is why it's all the more important to protect the cash system.

We are stuck in a global economy that feels as though it runs on autopilot. This generates a feeling of inevitability in which people believe that human society will become ever more automated and detached from the natural world, regardless of their desire. This feeling sparks many emotions, one of which is anxiety. My work on cash often brings me into contact with people who feel panicked at this sense of autopilot, and angry at their powerlessness in the face of it. One man sent me an email with a link to a CNBC story about how Visa is using Covid to promote a 'cashless Super Bowl'. The article noted how the company was loading the football stadium with gimmicks such as a 'reverse ATM' for people to convert their cash to digital money. There was only a single line in the email: 'bret so scary. can you stop this to save our cash'.

I do not know the man who sent this to me. He simply goes by the name AaronForTrump. This is not the first time a supporter of the controversial ex-US president has emailed me with a message of support for my pro-cash advocacy. I have also had flat-earthers and others who are deeply embroiled in conspiracy theories. Given that corporations have used the pandemic to extend their power even further, Covid and corporate capitalism are being fused in the eyes of conspiracy theorists. For example, my own content has been used in a documentary that claims Bill Gates was engineering Covid to push 5G and a cashless society to enslave everyone under an Orwellian New World Order. The video-makers found footage of me talking about the war on cash, in which I explain how it is driven by the numb expansion of the capitalist system, but that message was too subtle for them: they carefully selected only those sections of footage that would enable them to claim Bill Gates was personally orchestrating the drive against cash.

So, I found myself featured alongside far-right cranks and an evangelical preacher raging about a Satanic plot orchestrated by global elites. That video reached me via a WhatsApp message sent by a distant friend of my mother in South Africa. It was racing through her social media networks and reached millions of views before YouTube took it down. But this style of analysis – in which the disorientating effects of our mega-systems are attributed to single people like Bill Gates (or groups like 'the Jews', 'the Marxists' or 'the Elites') – is very popular. The global economy operates at a scale that few people have the time – and training – to make sense of, but people nevertheless must find narratives to explain the tremors of anxiety and instability that course through it (and by extension, them), and that feeling of autopilot. Conspiracy theorists step in to weave these narratives, and often align with the far-right in the process.

Cash now finds itself cast in these narratives, and not many in mainstream institutions are aware of this. In the early months of the pandemic I found myself invited to a private Facebook group called 'Keep Cash UK', which in the space of a few weeks had attracted over 30,000 members from all walks of life, vigorously vowing to challenge the corporate push against cash, while regularly promoting the latest conspiracy theory. The theories on the group were often crude, but were clearly emerging to make sense of intuitive concerns. The conspiracy theorists are wrong that individuals like Gates are single-handedly orchestrating a war on cash, but they are correct in their belief that the cash system stands in the way of the growing web of digital corporates who do not have our interests at heart. And the refusal of mainstream institutions to take this seriously is helping those virulent strains of far-right populism take root. Big institutions are more likely to ask Visa and Google to help them promote digital financial inclusion than they are to protect and promote the cash infrastructure that already exists. The fight against a global tech-finance vortex is thus in danger of falling under the banner of the far-right, when in reality it is all of us who should be concerned about this, regardless of our political orientation.

So, what do we do? Well one answer is that we must vigorously assert our right to use cash, and to see that as a political act. The desire to use cash is presented by the fintech industry as a stubborn refusal to move forward with the times, but I prefer to cast it as a stubborn refusal to go with the grain of corporate capitalism. We need to protect it, if for no other reason than to prove to ourselves that we are not mere avatars channelling the expansionary logic of a system indifferent to the people who constitute it. The pro-cash movement has a simple agenda of maintaining the existing peripheral cash system in the face of encroaching central digital ones, and while this will not lead to some utopian society, it can at least hold off a dystopian one.

My arguments for the protection of cash have been practical and political, but deep down I am fighting for something personal. The right to be dirty and physical. We are not immortal super-beings with our brains plugged into an AI cloud-complex, and neither would we want to be. We are messy and contradictory, and that is a spirit better protected by a more down-to-earth incarnation of money. Cash is credit money in a commodity body, birthed by large institutions but roaming among us like a gregarious friend to strangers. It is capitalist while preventing the expansion of capitalism. It is, in a word, syncretic. To use it one must accept slowness and contradiction, which – in a frenetic polarised world – is akin to a meditative practice.

Acknowledgements

Thanks to my incredible agent Patrick Walsh of PEW Literary, along with Margaret Halton and John Ash. Watching you guys work wonders has been a pleasure.

Thanks to the editorial heroes of the book, Stuart Williams and Lauren Howard of The Bodley Head, Hollis Heimbouch of Harper Business, and Ana Fletcher. You not only polished a rough diamond, but also extracted it from the ore. Thanks also to Daniel Halpern for believing in the book's vision. Much gratitude goes to all the other behind-the-scenes helpers at Penguin Random House and Harper Business.

A huge thanks to those who read the manuscript in order to provide editorial suggestions, technical input and corrections. This includes Rohan Grey, Sarah Jaffe, Phoebe Braithwaite, Victor Fleurot, Frederike Kaltheuner, Irene Claeys, Jules Porter and Tarn Rodgers John. Thanks for spurring me on with your encouraging words. Thanks also to Guillaume Lepecq for providing data.

Enormous thanks to the network of friends who kept me going through the extensive dark times. This includes the Floddenites (all of you), Julio Linares, Kei Kreutler, Eli Gothill, Jaya Klara Brekke, Matthew Lloyd, Dan Nixon, Lynne Davis, Adrian Blount, J.P. Crowe, Sam Gill, Rita Issa, Alistair Alexander, James Jackson, Saara Rei, Jules Mueller, Shaun Chamberlin, Alice Thwaite, Max Haiven, Cassie Thornton, Phoebe Tickell, Jutta Steiner, Monika Bielskyte, Stacco and Ann Marie, Nathaniel Calhoun, Steve Grumbine, Simon Youel, Joel Benjamin, Simka and Manu, and Glen Scott.

Special thanks to my family, who held me from afar.

Special thanks to Scott Lye and Jeff Cavaliere for keeping my body intact amidst the writing, and to Jürgen Carlo Schmidt for helping me to keep my soul intact.

Special thanks to Sophia, who inspired the proposal.

My respects to the late David Graeber, who passed away before I could ask him to endorse the book. Thanks for inspiring us.

Finally, a special thanks to the earth, for continuing to roll into the rays of the sun, and for providing a home for us all.

Notes

1. The Nervous System

p.24   *worked with the Berlin-based open data company OpenOil*: See 'How complex is BP? 1,180 companies across 84 jurisdictions going 12 layers deep', 3 Sept. 2014,https://blog.opencorporates.com/2014/09/03/how-complex-is-bp-1180-companies-across-84-jurisdictions-going-12-layers-deep/

p.24   *up to half of global trade actually takes place* within *corporations*: Paweł Folfas, 'Intra-firm trade and non-trade intercompany transactions: changes in volume and structure during 1990–2007', p.3, https://www.etsg.org/ETSG2009/papers/folfas.pdf

2. The War on Cash

p.32   *5.2 million payments attempts were blocked*: These details can be found in Visa UK's explanatory letter to the UK Parliament on 15 June 2018, accessible at https://www.parliament.uk/globalassets/documents/commons-committees/treasury/Correspondence/2017-19/visa-response-150618.pdf

p.32   *the Bangladeshi Central Bank's account at the US Federal Reserve*: An overview of this heist can be found at https://en.wikipedia.org/wiki/Bangladesh_Bank_robbery

p.34   *central bank research showed that the PIN pads associated with digital payment pose a greater risk*: Bank of England, 'Cash in the time of Covid', *Quarterly Bulletin* 2020 Q4, 24 Nov. 2020, https://www.bankofengland.co.uk/quarterly-bulletin/2020/2020-q4/cash-in-the-time-of-covid

p.35   *'Reports on the End of Cash are Greatly Exaggerated'*: Federal Reserve Bank of San Francisco, 20 Nov. 2017, https://www.frbsf.org/our-district/about/sf-fed-blog/reports-death-of-cash-greatly-exaggerated/

p.36   *central banks recorded a large increase in cash withdrawn from ATMs*: See for example, 'Cash still king in times of COVID-19' by the ECB's Fabio Panetta https://www.ecb.europa.eu/press/key/date/2021/html/ecb.sp210615~05b32c4e55.en.html, the Bank of England's 'Cash in the time of Covid', https://www.bankofengland.co.uk/quarterly-bulletin/2020/2020-q4/cash-in-the-time-of-covid and the Federal Reserve's '2021 findings from the Diary of Consumer Payment Choice' https://www.frbsf.org/cash/publications/fed-notes/2021/may/2021-findings-from-the-diary-of-consumer-payment-choice/

p.36   *including those who don't want their wealth locked up in banks during a banking crisis*: Morten Bech, Umar Faruqui, Frederik Ougaard & Cristina Picillo, 'Payments are a-changin' but cash still rules', *BIS Quarterly Review*, March 2018, https://www.bis.org/publ/qtrpdf/r_qt1803g.pdf

p.36   *The Federal Reserve sees huge increases in cash demand prior to hurricanes*: This was communicated during a private presentation by Alex Bau of the Federal Reserve's Cash Product Office, titled 'Understanding Cash Usage: Rethinking Volume Forecasting', Feb. 2019

p.36   *it accounted for over 50 per cent of transactions under $10 and 30 per cent of overall payments volume*: See Federal Reserve Bank of San Francisco, '2018 Findings from the Diary of Consumer Payment Choice' 15 Nov. 2018, https://www.frbsf.org/cash/files/federal-reserve-cpo-2018-diary-of-consumer-payment-choice-110118.pdf

p.38   *'we want a cashless society'*: Javier E. David, 'Bank of America CEO: "We want a cashless society"', *Yahoo!Finance*, 19 June 2019, https://finance.yahoo.com/news/bank-of-america-brian-moynihan-cashless-society-210717673.html

p.39   *Banks try all manner of tactics to discourage cash usage*: See for example, Risen Jayaseelan, 'The joys and sufferings of going cashless', *The Star*, 7 Jan. 2019, https://www.thestar.com.my/business/business-news/2019/01/07/the-joys-and-sufferings-of-going-cashless/

p.40   *'cashless man of India' campaign, and another under the banner that 'Kindness is Cashless'*: For details of these campaigns, see https://www.visa.com.bz/visa-everywhere/global-impact/cashless-man-of-india.html and https://www.campaignindia.in/video/visa-looks-to-spread-goodness-and-education-with-kindnessiscashless/434240

p.40   *the 'campaign is the latest step of Visa UK's long-term strategy to make cash "peculiar" by 2020'*: Visa UK has now removed this reference from their website, but a version of it can be found on the Internet Archive here https://web.archive.org/web/20171016092002/https://www.visa.co.uk/newsroom/visa-europe-launches-cashfree-and-proud-campaign-1386958?returnUrl=%2fnewsroom%2fcash-free-and-proud-video-female-22806.aspx

p.40   *handed out $10,000 prizes to small trendy businesses that 'go cashless'*: See 'Meet the Cashless Challenge winners' at https://usa.visa.com/visa-everywhere/innovation/visa-cashless-challenge-winners.html

p.41   *Payments companies even spread ideas about cash as environmentally unsustainable*: For an example of this, see 'Why Going Cashless Is Better for the Environment', written by digital payments company Pomelo Pay https://www.pomelopay.com/blog/cashless-better-for-environment

p.41   *Amazon lobbied against legislation requiring shops to accept cash*: The *Philadelphia Inquirer* followed this story. See 'Amazon warns it may rethink plans to open a Philly store if the city bans cashless retailers', 15 Feb. 2019, https://www.inquirer.com/business/retail/amazon-go-philadelphia-cashless-store-ban-20190215.html, and 'Emails show how Philly officials tried to help Amazon escape proposed cashless store ban', 26 Feb. 2019, https://www.inquirer.com/news/amazon-go-cashless-store-philadelphia-lobbying-20190226.html

p.42   *'cash thresholds' to prevent the use of cash over a certain amount*: For a review of cash thresholds, see Peter Sands, Haylea Campbell, Tom Keatinge and Ben

Weisman, 'Limiting the Use of Cash for Big Purchases: Assessing the Case for Uniform Cash Thresholds', M-RCBG Associate Working Paper Series No. 80, Sept. 2017, https://www.hks.harvard.edu/sites/default/files/centers/mrcbg/files/80_limiting.cash.pdf

p.42 *recently proposed to ratchet it down further to €300*: Guillaume Lepecq, 'ECB calls for Greece to drop Cash Payment Limitations. But what about the other Countries?', *CashEssentials*, 3 Dec. 2019, https://cashessentials.org/ecb-calls-for-greece-to-drop-cash-payment-limitations-but-what-about-the-other-countries/

p.43 *the 2020 Corruption Perceptions Index*: The index can be found here https://www.transparency.org/en/cpi/2020

p.43 *it is often coupled with a refusal to accept state cash for state services*: There are many examples of this, but perhaps the most illuminating case to follow is Norbert Häring's campaign to pay for his public radio licence in cash in Germany, which led to the European Court of Justice getting involved. See the timeline here https://norberthaering.de/en/my-ecj-courtcase-on-cash/timeline/

p.44 *'Paytm congratulates Honorable Prime Minister Sh. Narendra Modi on taking the boldest decision in the financial history of Independent India!'*: For images of these advertisements see, 'E-commerce companies are bombarding us with front page ads after demonetisation', *The News Minute*, 10 Nov. 2016, https://www.thenewsminute.com/article/e-commerce-companies-are-bombarding-us-front-page-ads-after-demonetisation-52670

p.45 *Malaysia, for example, has reputedly explored demonetisation*: These rumours began to spread after the Malaysian prime minister Najib Razak praised Narendra Modi as a 'good reformist' after his 2016 demonetisation effort.

p.47 *The German Bundesbank is notable for its support of cash*: Central bank employees are generally discouraged from overtly promoting cash, but the Bundesbank often posts reports that cast a favourable light on it. This includes research that challenges the idea that cash is mostly used for the shadow economy (see 'Cash demand in the shadow economy', Deutsche Bundesbank Monthly Report March 2019, https://www.bundesbank.de/resource/blob/793190/466691bce4f27f76407b35f8429441ae/mL/2019-03-bargeld-data.pdf) and research that challenges the idea of cash being slow and inconvenient (see 'Study finds that cash payments are quick and cheap', 12 Feb. 2019, https://www.bundesbank.de/en/tasks/topics/study-finds-that-cash-payments-are-quick-and-cheap-776688). The Bundesbank was among the first to counter claims that cash spread COVID-19 (see 'Cash poses no particular risk of infection for public', 18 March 2020, https://www.bundesbank.de/en/tasks/topics/cash-poses-no-particular-risk-of-infection-for-public-828762

p.47 *Along with researchers at the IMF*: See for example, Katrin Assenmacher & Signe Krogstrup, 'Monetary Policy with Negative Interest Rates: Decoupling Cash from Electronic Money', IMF Working Paper No. 18/191, 27 Aug. 2018, https://www.imf.org/en/Publications/WP/Issues/2018/08/27/Monetary-Policy-with-Negative-Interest-Rates-Decoupling-Cash-from-Electronic-Money-46076

p.48 *Dutch central banks voicing reservations about its disappearance*: See for example, 'Dutch central bank concerned about decreasing use of cash', *NL Times*, 29 Oct. 2018, https://nltimes.nl/2018/10/29/dutch-central-bank-concerned-decreasing-use-cash and 'Cash is still king: central bank calls for action to keep notes and coins', *DutchNews*, 7 July 2021, https://www.dutchnews.nl/news/2021/07/cash-is-still-king-central-bank-calls-for-action-to-keep-notes-and-coins/

p.48   *'If Crisis or War Comes'*: The Swedish information booklet can be found here https://www.dinsakerhet.se/siteassets/dinsakerhet.se/broschyren-om-krisen-eller-kriget-kommer/om-krisen-eller-kriget-kommer---engelska.pdf

3. The Giant in the Mountain

p.60   *counterfeits have even been used as political weapons*: See Karl Rhodes, 'The Counterfeiting Weapon', *Econ Focus, Federal Reserve Bank of Richmond*, vol. 16 (1Q), pages 34–37, https://www.richmondfed.org/-/media/richmondfedorg/publications/research/econ_focus/2012/q1/pdf/economic_history.pdf

p.60   *In contemporary times Indian politicians accuse Pakistan's secret services:* See for example Shiban Khaibri 'Fighting economic terrorism', *Daily Excelsior*, 19 Oct. 2013, https://www.dailyexcelsior.com/fighting-economic-terrorism/

4. Digital Chips

p.66   *In Romania, for example, 86 per cent of people have bank accounts*: 'Statistics show gap between Romanians' card ownership and usage', *Romania Insider*, 24 June 2019, https://www.romania-insider.com/lidl-card-usage-2019

p.77   *'along the Eastern seaboard'*: Tony Kontzer, 'Inside Visa's Data Center', *Network Computing*, 29 May 2013, https://www.networkcomputing.com/networking/inside-visas-data-center

p.79   *in 2013 the European Central Bank set up a 'swap line' with the People's Bank of China*: European Central Bank, 'ECB and the People's Bank of China establish a bilateral currency swap agreement', 10 Oct. 2013 https://www.ecb.europa.eu/press/pr/date/2013/html/pr131010.en.html

5. The Bank-Chip Society

p.83   *number of UK bank branches declined by 28 per cent, while ATM numbers declined by 24 per cent*: UK Parliament, 'Statistics on access to cash, bank branches and ATMs', 12 Oct. 2021 https://commonslibrary.parliament.uk/research-briefings/cbp-8570/

p.84   *Worldwide Google Trends data shows a significant increase in searches for the term 'cashless'*: Find the interactive chart here https://trends.google.com/trends/explore?date=all&q=cashless

p.86   *'Cyber Monday', for example, is no more than a creation of the National Retail Federation*: Jessika Toothman & Kathryn Whitbourne, 'How Cyber Monday Works', *HowStuffWorks*, 2 Dec. 2019, https://money.howstuffworks.com/personal-finance/budgeting/cyber-monday1.htm

p.92   *Data from several studies shows that cash usage is lowest among those with higher incomes*: See for example John Bagnall, David Bounie, Kim P. Huynh, Anneke Kosse, Tobias Schmidt, Scott Schuh & Helmut Stix, 'Consumer Cash Usage: A Cross-Country Comparison with Payment Diary Survey Data', *European Central Bank Working Paper Series*, No. 1685, June 2014, pp.18–19, https://www.ecb.europa.eu/pub/pdf/scpwps/ecbwp1685.pdf

p.93   *publicly praised India's Modi government for clamping down on cash*: On 8 November 2018 Thaler commented on Modi's decision, tweeting, 'This is a policy I have long supported. First step toward cashless and good start on reducing corruption', https://twitter.com/r_thaler/status/796007237458206720

6. Big Brother. Big Bouncer. Big Butler

p.109   *PayPal is entitled to share your data with 600 organisations*: Wolfie Christl, 'Corporate surveillance in everyday life: How companies collect, combine, analyze, trade, and use personal data on billions', *Cracked Labs* report, June 2017, p.21

p.109   *Google, for example, utilises sources like this, claiming that*: Sridhar Ramaswamy, 'Powering ads and analytics innovations with machine learning', *Google Inside AdWords*, 23 May 2017, https://adwords.googleblog.com/2017/05/powering-ads-and-analytics-innovations.html

p.109   *Bloomberg reported that Google entered into a secret deal to purchase credit card data*: Mark Bergen & Jennifer Surane, *Bloomberg*, 'Google and Mastercard Cut a Secret Ad Deal to Track Retail Sales', 30 Aug. 2018, https://www.bloomberg.com/news/articles/2018-08-30/google-and-mastercard-cut-a-secret-ad-deal-to-track-retail-sales

p.109   *banks like Wells Fargo and Citi build profiles of their customers' spending behaviour*: See for example, Blake Ellis, 'The banks' billion-dollar idea', *CNN Money*, 8 July 2011, https://money.cnn.com/2011/07/06/pf/banks_sell_shopping_data/

p.110   *Two months later the UK Taylor Review called for a crackdown on the use of cash*: The July 2017 Taylor Review of Modern Working Practices can be found here https://assets.publishing.service.gov.uk/government/uploads/system/uploads/attachment_data/file/627671/good-work-taylor-review-modern-working-practices-rg.pdf. See also Vanessa Holder, 'Crackdown proposed on cash-in-hand payments', *Financial Times*, 12 July 2017, https://www.ft.com/content/c215174a-6640-11e7-9a66-93fb352ba1fe

p.110   *'A core responsibility of the IMF is to oversee the international monetary system'*: See IMF Surveillance factsheet https://www.imf.org/en/About/Factsheets/IMF-Surveillance

p.111   *Section 314 (a) of the Act allows targeted individuals' accounts to be monitored*: See FinCEN's 314(a) Fact Sheet https://www.fincen.gov/sites/default/files/shared/314afactsheet.pdf

p.111   *In 2010 an FBI document was leaked describing their 'hotwatch' system*: Ryan Singel, 'Feds Warrantlessly Tracking Americans' Credit Cards in Real Time', *Wired*, 2 Dec. 2010

p.112   Der Spiegel *released an exclusive about the National Security Agency's 'Follow the Money'*: 'NSA Spies on International Payments', *Der Spiegel*, 15 Sept. 2013 https://www.spiegel.de/international/world/spiegel-exclusive-nsa-spies-on-international-bank-transactions-a-922276.html

p.112   *The SWIFT network is also subject to this spying*: For details on SWIFT surveillance and its legality, see Johannes Köppel, *The SWIFT Affair: Swiss Banking Secrecy and the Fight against Terrorist Financing*. New edition [online]. Genève: Graduate Institute Publications, 2011 (generated 03 Nov. 2021). DOI: https://doi.org/10.4000/books.iheid.225

p.113 *The scholar Nathalie Maréchal chronicles this in her academic paper*: Nathalie Maréchal, 'First They Came for the Poor: Surveillance of Welfare Recipients as an Uncontested Practice', *Media and Communication*, Vol. 3, No. 3, 20 Oct. 2015 https://doi.org/10.17645/mac.v3i3.268

p.113 *their spending via pocket money apps*: For an example of this see the GoHenry app https://www.gohenry.com/uk/benefits-for-parents/

p.115 *There are now many reports of Chinese people being blocked from travel and other 'privileges'*: See Harry Cockburn, 'China blacklists millions of people from booking flights as "social credit" system introduced', *Independent*, 22 Nov. 2018. See also Ed Jefferson, 'No, China isn't Black Mirror – social credit scores are more complex and sinister than that', *New Statesman,* 27 April 2018

p.116 *It managed to win a temporary injunction against Barclays*: See Dahabshiil's press release about the injunction here https://www.dahabshiil.com/blog/dahabshiilwins-injunction-against-barclays-1/

7. The Unnatural Progress of a 'Rapidly Changing World'

p.121 *because over half will no longer deal in cash*: Josh Robbins, 'Bank and ATM closures: what the UK can learn from Sweden', *Which?*, 20 Dec. 2019, https://www.which.co.uk/news/2019/12/bank-and-atm-closures-what-the-uk-can-learn-from-sweden/

p.127 *one major reason people cite for preferring cash is its tangible physicality*: See for example, Deutsche Bundesbank, 'Payment behaviour in Germany in 2017', https://www.bundesbank.de/resource/blob/737278/458ccd8a8367fe8b36bb fb501b5404c9/mL/zahlungsverhalten-in-deutschland-2017-data.pdf and Chris Jennings, 'Survey: It's a Card-Obsessed World, but Cash Is Still King—Here's Why', *GoBankingRates*, 1 Nov. 2019 https://www.gobankingrates.com/credit-cards/advice/survey-americans-prefer-cash-to-credit/

p.127 *Several studies have shown that digital payments encourage fast and detached spending*: See for example, Martina Eschelbach, 'Pay cash, buy less trash? – Evidence from German payment diary data', *International Cash Conference 2017*, https://www.econstor.eu/handle/10419/162908 and Drazen Prelec & Duncan Simester, 'Always Leave Home Without It: A Further Investigation of the Credit-Card Effect on Willingness to Pay', *Marketing Letters*, Vol. 12, 2001, pp.5–12 https://link.springer.com/article/10.1023/A:1008196717017. For a shorter overview, see Hal E. Hershfield, 'The Way We Spend Impacts How We Spend', *Psychology Today*, 10 July 2012, https://www.psychologytoday.com/us/blog/the-edge-choice/201207/the-way-we-spend-impacts-how-we-spend

p.127 *in its 'benefits of going cashless' website, Visa reports*: See 'Are cashless payments good for business?' https://usa.visa.com/visa-everywhere/innovation/benefits-of-going-cashless.html

p.127 *to which end it partners with – among others – Visa and Google*: USAID has removed this document from its original location, but for an archived version see https://web.archive.org/web/20201016190208/https://www.usaid.gov/sites/default/files/documents/15396/Lab-Fact-Sheet.pdf

p.128 *It has a manual about how to digitise the Indian monetary system*: This document, originally titled 'Beyond Cash', has since been removed from the Internet, but an archived version can be found here https://web.archive.org/

web/20210731050518/https://www.digitaldevelopment.org/beyond-cash. For a broader glimpse into USAID's strategy for advancing digital payments across the world, see its 'Mission Critical: Enabling Digital Payments for Development' briefing, https://www.usaid.gov/sites/default/files/documents/15396/USAID-DFS-OpportunityBrief.pdf

p.133 *Tellingly, however, stories surfaced about Amazon doing behind-the-scenes lobbying*: See 'Amazon warns it may rethink plans to open a Philly store if the city bans cashless retailers', 15 Feb. 2019, https://www.inquirer.com/business/retail/amazon-go-philadelphia-cashless-store-ban-20190215.html and 'Emails show how Philly officials tried to help Amazon escape proposed cashless store ban', 26 Feb. 2019, https://www.inquirer.com/news/amazon-go-cashless-store-philadelphia-lobbying-20190226.html

p.133 *The* Wall Street Journal *questioned whether the legislation placed 'limits on innovation'*: Scott Calvert, 'Philadelphia Is First U.S. City to Ban Cashless Stores', *Wall Street Journal*, 7 March 2019, https://www.wsj.com/articles/philadelphia-is-first-u-s-city-to-ban-cashless-stores-11551967201

8. Shedding and Re-skinning

p.143 *Commerzbank has set up its own fintech incubator*: See Commerzbank's description here https://www.commerzbank.de/en/nachhaltigkeit/markt__kunden/mittelstand/main_incubator/main_incubator.html

p.144 *Mel Evans refers to this phenomenon by which coldly commercial institutions patronise art as 'artwash'*: Mel Evans, *Artwash: Big Oil and the Arts* (Pluto Press, 2015)

p.144 *a Kentridge video animation called 'Second Hand Reading'*: The video can be seen here https://www.youtube.com/watch?v=IEfUjg5viGk

p.147 *has a digital interface with a human first name – Erica*: See Bank of America's introduction to Erica here https://promotions.bankofamerica.com/digital-banking/mobilebanking/erica

p.147 *HSBC, for example, is branding the 'sound of HSBC' for its chatbots*: See HSBC's Josh Bottomley comment on this at CogX 2019 here https://youtu.be/lqnZVXCRhZM?t=1620

p.150 *Alipay and WeChat in China have pioneered payment by facial recognition*: 'Smile-to-pay: Chinese shoppers turn to facial payment technology', *Guardian*, 4 Sept. 2019, https://www.theguardian.com/world/2019/sep/04/smile-to-pay-chinese-shoppers-turn-to-facial-payment-technology

9. Sherlock Holmes and the Strange Case of the Data Ghost

p.154 *Christian evangelist Meghan O'Gieblyn argues that Singularity stories*: Meghan O'Gieblyn, 'God in the machine: my strange journey into transhumanism', *Guardian*, 18 April 2017, https://www.theguardian.com/technology/2017/apr/18/god-in-the-machine-my-strange-journey-into-transhumanism

p.155 *Thiel then funded Lonsdale and a few others to apply the same model to national security*: See Peter Waldman, Lizette Chapman & Jordan Robertson, 'Palantir Knows Everything About You', *Bloomberg Businessweek*, 19 April 2018, https://www.bloomberg.com/features/2018-palantir-peter-thiel/

p.156 *Palantir rents itself out to major financial corporations like J. P. Morgan*: Waldman et al (see above).

p.156 *HSBC's interactions with its 39 million customers, for example, have generated some 150 petabytes of data*: See HSBC's Josh Bottomley discuss these figures at CogX 2019 https://youtu.be/boxy3llT-00?t=423

p.157 *The Royal Bank of Canada, for example, has a team of 100 PhDs researching AI*: See Foteini Agrafioti, Head of RBC's Borealis AI, discuss the team here https://www.rbccm.com/en/insights/story.page?dcr=templatedata/article/insights/data/2019/04/pushing_the_boundaries_of_science_with_ai

p.157 *Now we've got three, and it seems quite natural to us that evolution*: Joanne Hannaford of Goldman Sachs talking at CogX 2019 https://youtu.be/lqnZVXCRhZM?t=749

p.167 *'Loan denial rates in the US are historically far higher for black people'*: See for example, Sray Agrawal, 'Fair AI: How to Detect and Remove Bias from Financial Services AI Models', *Finextra*, 11 Sept. 2019, https://www.finextra.com/blogposting/17864/fair-ai-how-to-detect-and-remove-bias-from-financial-services-ai-models

p.168 *Privacy International reported in 2015 that the Ugandan government was pioneering all manner of surveillance techniques*: Privacy International, *For God and My President: State Surveillance In Uganda*, Oct. 2015, https://privacyinternational.org/sites/default/files/2017-12/Uganda_Report_1.pdf

p.168 *the overall paradigm has often led to over-indebtedness*: See Isabelle Guérin, Solène Morvant-Roux, Magdalena Villarreal (editors), *Microfinance, Debt and Over-Indebtedness Juggling with Money* (Routledge, 2014)

p.169 *Lenddo, which analyse the 'digital footprint' left by someone's phone usage*: See for example, Privacy International, 'Fintech's dirty little secret? Lenddo, Facebook and the challenge of identity', 23 Oct. 2018, https://privacyinternational.org/long-read/2323/fintechs-dirty-little-secret-lenddo-facebook-and-challenge-identity. See also María Óskarsdóttir, Cristián Bravo, Carlos Sarraute, Bart Baesens & Jan Vanthienen, 'Credit Scoring for Good: Enhancing Financial Inclusion with Smartphone-Based Microlending', *Thirty Ninth International Conference on Information Systems, San Francisco*, 2018, https://eprints.soton.ac.uk/425943/1/Credit_Scoring_for_Good_Enhancing_Financial_Inclusion_with_Smart.pdf.

p.169 *The company claims to use machine-learning to analyse up to 12,000 variables to work out a credit score*: Watch 'How does Lenddo work?' on the Omidyar Network's YouTube channel https://www.youtube.com/watch?v=0bEJO4Twgu4

p.169 *Companies like Tala in Kenya analyse 10,000 data points*: See Kate Douglas, 'How calling your mother can help you get a micro-loan in Kenya', *How we made it in Africa*, I Nov. 2016, https://www.howwemadeitinafrica.com/calling-mother-can-help-get-micro-loan-kenya/56525/

p.169 *Neener Analytics apply psychometric testing techniques to 'undecisionable' people*: See the Neener team's description of what they do here http://www.neener-analytics.com/what-we-do.html, and their Finovate presentation here https://finovate.com/videos/finovatespring-2019-neener-analytics/

p.172 *the sound of an automated debt-collection agent's voice we hear*: Voca.ai (now acquired by Snap) provides this service. See a demonstration here https://www.youtube.com/watch?v=FBcaCp8CObA

10. Clash of the Leviathans

p.179   *pressurised by by-laws like the US PATRIOT Act*: See for example, Eric J. Gouvin, 'Bringing Out the Big Guns: The USA PATRIOT Act, Money Laundering, and the War on Terrorism', 55 *Baylor L. Rev.* 955, 2003, and Fletcher N. Baldwin, 'Money laundering countermeasures with primary focus upon terrorism and the USA Patriot Act 2001', *Journal of Money Laundering Control*, Vol. 6 No. 2, 2002, pp.105–136

12. The Political Tribes of Cyber-Kowloon

p.215   *The horseshoe theory of politics*: For a basic overview of the theory, see https://en.wikipedia.org/wiki/Horseshoe_theory. Some versions of the theory see a tendency towards authoritarianism as being a common feature in both the far right and far left, but anti-statism can be a common feature too, alongside a common rejection of the centre of the political spectrum.

p.216   *The US Bitcoin trader Chad Elwartowski, for example, rigged up a seasteading dwelling*: Adam Forrest, 'US Bitcoin trader faces death penalty after Thai navy seizes floating home of fugitive "seasteaders"', 20 April 2019, https://www.independent.co.uk/news/world/asia/bitcoin-chad-elwartowski-thai-navy-floating-home-seasteading-phuket-a8878981.html

p.216   *'Bitcoin is what they fear it is, a way to leave . . . to make a choice'*: The statement comes from the original Dark Wallet promotional video, which can be found here https://www.youtube.com/watch?v=Ouo7Q6Cf_yc

p.220   *a term first used in 1994 by the cryptographer Nick Szabo*: Nick Szabo, 'The Idea of Smart Contracts', 1997, http://www.fon.hum.uva.nl/rob/Courses/InformationInSpeech/CDROM/Literature/LOTwinterschool2006/szabo.best.vwh.net/idea.html

p.225   *Steve Bannon believes in Bitcoin as a means to drive a 'global populist revolt'*: Billy Bambrough, 'Donald Trump And Steve Bannon In Surprise Bitcoin Split', 5 Aug.2019,https://www.forbes.com/sites/billybambrough/2019/08/05/donald-trump-and-steve-bannon-in-surprise-bitcoin-split/

13. Raiding the Raiders

p.230   *investment banks like Goldman Sachs began flirting with the idea of starting crypto trading divisions*: https://www.wsj.com/articles/goldman-sachs-explores-a-new-world-trading-bitcoin-1506959128 and https://www.bloomberg.com/news/articles/2017-10-02/goldman-sachs-said-to-explore-starting-bitcoin-trading-venture

p.232   *'blockchain' solutions for any form of interbank co-ordination, from securitisation*: See for example William Suberg, 'Blockchain Meets Securitization Market In New Chamber Of Digital Commerce Partnership', *Cointelegraph*, 1 March 2017, https://cointelegraph.com/news/blockchain-meets-securitization-market-in-new-chamber-of-digital-commerce-partnership

p.232   *A team from J. P. Morgan itself built its own DLT system called Quorum*: Quorum has now been acquired by ConsenSys https://consensys.net/quorum/

p.233   *NASDAQ has announced a partnership with R3*: See R3 press release 'Nasdaq to collaborate with R3 on institutional grade offerings for digital assets exchanges', 29 April 2020, https://www.r3.com/press-media/nasdaq-to-collaborate-with-r3-on-institutional-grade-offerings-for-digital-assets-exchanges/

p.235   *Tether tokens being partially 'unbacked'*: There is a large amount of reporting on the Tether controversy, but for an illustrative piece, see Nikhilesh De, 'Tether Says Its Stablecoin Is "Fully Backed" Again' *Coindesk*, 8 Nov. 2019, https://www.coindesk.com/tether-says-its-stablecoin-is-fully-backed-again

p.235   *to which Tether's US dollar reserves could be moved*: See Amy Castor, 'The curious case of Tether: a complete timeline of events', 17 Jan. 2019, https://amycastor.com/2019/01/17/the-curious-case-of-tether-a-complete-timeline-of-events/

p.235   *crypto-dollar called USD Coin to you in return, via an Ethereum smart contract*: See the FAQ on USD Coin here https://help.coinbase.com/en/coinbase/getting-started/crypto-education/usd-coin-usdc-faq

p.238   *headlines claiming that Libra would 'destroy commercial and central banks'*: Robert Johnson, 'BitMEX CEO: "Libra Will Destroy Commercial & Central Banks"', 2 July 2019, https://cryptodaily.co.uk/2019/07/bitmex-ceo-libra-will-destroy-commercial-central-banks

p.241   *arguing that Libra would be a US counterforce to Chinese digital payments platforms*: read the testimony here https://financialservices.house.gov/uploadedfiles/hhrg-116-ba00-wstate-marcusd-20190717.pdf

p.241   *a message that Facebook CEO Mark Zuckerberg pushed too*: Josh Constine, 'Facebook's regulation dodge: Let us, or China will', *TechCrunch*, 17 July 2019, https://techcrunch.com/2019/07/17/facebook-or-china/

p.244   *One major central bank think-tank surveyed twenty-three central banks*: OMFIF, *Retail CBDCs: The next payments frontier*, https://www.omfif.org/ibm19/

Conclusion

p.249   *In the UK this has worked, with cash usage plummeting by over 50 per cent in 2020*: This figure is based on Enryo research. See https://enryo.org/news-%26-media/f/cash-usage-falls-by-over-50%25-but-will-remain-stable-until-2030

p.249   *in 2020 Amazon's revenues rose by over 40 per cent*: Tom Huddleston Jr., 'How much revenue tech giants like Amazon and Apple make per minute', *CNBC*, 1 May 2021, https://www.cnbc.com/2021/05/01/how-much-revenue-tech-giants-like-amazon-and-apple-make-per-minute.html

Index

About the Author

BRETT SCOTT is an economic anthropologist, financial activist, and former broker. In 2013 he published *The Heretic's Guide to Global Finance: Hacking the Future of Money*, and since then has spoken at hundreds of events across the globe and has appeared across international media, including BBC World News and Sky News. He has written extensively on financial reform, digital finance, alternative currency, blockchain technology, and the cashless society for publications like the *Guardian*, *New Scientist*, *Huffington Post*, *Wired*, and CNN.com, and also publishes the *Altered States of Monetary Consciousness* newsletter. He has worked on financial reform campaigns and alternative currency systems with a wide range of groups and is a Senior Fellow of the Finance Innovation Lab (UK). He lives in Berlin.